Human-Plant Entanglement: Thinking with Plants in the Anthropocene

Critical Plant Studies

PHILOSOPHY, LITERATURE, CULTURE

Series Editor

Michael Marder
(*IKERBASQUE/The University of the Basque Country, Vitoria*)

VOLUME 9

The titles published in this series are listed at *brill.com/cpst*

Human-Plant Entanglement: Thinking with Plants in the Anthropocene

Edited by

Ratul Nandi
Jagannath Basu
Jayjit Sarkar

BRILL

LEIDEN | BOSTON

Cover image: Adrien King (2019), available via http://unsplash.com.

Library of Congress Cataloging-in-Publication Data

Names: Nandi, Ratul, editor. | Basu, Jagannath, 1987– editor. | Sarkar, Jayjit, editor.
Title: Human-plant entanglement : thinking with plants in the anthropocene / edited by Ratul Nandi, Jagannath Basu, Jayjit Sarkar.
Description: Leiden ; Boston : Brill, [2025] | Series: Critical plant studies : philosophy, literature, culture, 2213-0659 ; volume 9 | Includes bibliographical references and index. | Summary: "Human-Plant Entanglement: Thinking with Plants in the Anthropocene is an edited collection that redefines the boundaries of phytocentric scholarship. By foregrounding the question of the Anthropocene at the centre of plant studies, this book illustrates how attentiveness to plant life can allow our habitual anthropocentric/instrumental assumptions to be invaded by a unique 'phytocentric' impression that presents a new ethical imaginary for a human-plant relationship. With twelve carefully argued essays, this book sets a new benchmark in the field of Critical Plant Studies"— Provided by publisher.
Identifiers: LCCN 2024057518 (print) | LCCN 2024057519 (ebook) | ISBN 9789004720633 (hardback) | ISBN 9789004721777 (e-book)
Subjects: LCSH: Plants (Philosophy) | Human-plant relationships. | Environmental ethics.
Classification: LCC B105.P535 H86 2025 (print) | LCC B105.P535 (ebook) | DDC 113/.8—dc23/eng/20241214
LC record available at https://lccn.loc.gov/2024057518
LC ebook record available at https://lccn.loc.gov/2024057519

Typeface for the Latin, Greek, and Cyrillic scripts: "Brill". See and download: brill.com/brill-typeface.

ISSN 2213-0659
ISBN 978-90-04-72063-3 (hardback)
ISBN 978-90-04-72177-7 (e-book)
DOI 10.1163/9789004721777

This book is printed on acid-free paper and produced in a sustainable manner.

Contents

Foreword

1 Do Trees Laugh?

When I first read about plants 'crying' when 'water-deprived' or 'injured' in early spring this year, I thought of two things. The first was Jagadish Chandra Bose, the scientist working in his lab in colonial Calcutta a hundred years ago, designing instruments to record plant response to various kinds of external stimuli, and using a register of emotions reserved almost exclusively for humans. Both these actions were revolutionary: the desire to record what Bose called the plant's handwriting, but, even more importantly, for a scientist to reject the adopted dry voice of science journals for a humane vocabulary of thrill, desire, wonder and love. The second had to do with a joke in my family: how can someone like me—whose strongest pull is for humour, for people who make me laugh—be so attracted to plant life, living beings without humour?

Jagadish Ch. Bose, interested in understanding the 'language' of plants, recorded their responses to 'external stimuli', particularly to 'injury'. The 'torulipi'—plant script—that his instruments recorded were, quite obviously, of a plant under stress. If recent research proves Bose's deductions from more than a century ago, that plants 'cry', does it leave out the possibility of the existence of laughter in the plant world?

And yet, humans seem to have heard trees laughing. This is from Amit Chaudhuri's song "Laughter", written and composed in the late 1970s, when he was in his late teens.

I heard the laughter of the trees
I heard them laughing endlessly
I walked with my eyes hanging down
I thought they were laughing at me

Was it a dream in which I heard
All in my mind all in my mind
And though I walked on I could not
Leave my loneliness behind

Who can forget the laughter?
Who can forget the trees?
Who who saw them swaying
Laughing laughing at me?

And as I walked I saw the leaves
Fall like dying stars below
Across my eyes across my soul
I felt a wind begin to blow

And all was silent like the sky
And all was still just as the sea
But only only as I cried
The trees were laughing endlessly

I've always been moved by the song, moved by its beauty as much as by the awareness of a teenager writing this. Not until I had heard this song did the question occur to me—is it possible for living beings without eyes to respond with laughter? This song is neither personification nor pathetic fallacy, two easy ways adopted to write about those with whom we do not share a language. This takes the laughter of trees to be as natural as the laughter of humans. By this I mean that it does not imagine a tree to be a person who is laughing but that all living beings have access to laughter. Could this be true, that everything that breathes, that has a living relationship with air, can laugh? It is at this moment that something integral to my consciousness rises to the surface—my life of thought with the rasa theory, a theory of art emotions that goes back to Bharata's Natya Shastra. Bharata, basing his theory on the plays that he had watched, particularly those written by Kalidasa, identified eight rasas: shringar, hasya, raudra, karuna, veera, bhayanaka, bibhatsa, adbhuta. The Kashmiri thinker and aesthete Abhinavagupta added a ninth about five centuries later: shanta, peace. When I teach the rasa theory to my students, we often discuss their human-centredness, we wonder whether we could imagine the same set of emotions and responses in a world beyond humans. Could they hold true for plant life, for instance? Jagadish Bose was mocked for his claims about plant sentience, that plants do feel—it was, after all, this 'feeling' that he wanted to record. There's also a binary in Chaudhuri's song about the laughing trees that is bound to trouble us emotionally—as the 'I' cries, the trees 'laugh endlessly'. Could that be possible, this cruel laughter of plant life? As I run through the other (art)-emotions on the list, I think of scientists in the last one hundred years since Bose having worked towards proving that plants do indeed 'feel' many of these feelings—the ones related to stress and, consequently, an impending death.

But laughter? Why does laughter—hasya—feel so alien to the plant world?

The leaves of the touch-me-not fold when touched, but it generates wonder, not laughter. But take what locals in the Indian state of Uttarakhand have been calling a 'gudgudi'—tickle—tree. You tickle the tree's trunk like you tickle a child, and the leaves on the branches begin to move. 'At first you think that it might be the wind tickling it,' says the environmental journalist from whose video I first learn about such a tree in Katarniaghat, 'Then you realise that it's not the air, for the leaves of no other tree in its surroundings are moving'. *This* is the thing—that the tree's response to the air, as in the shiver of the leaves, does not surprise us, but its response to tickling does. Such is our thrill at this discovery of the laughter of trees that this discovery comes to us in the form of a news report, the narrative as exclamatory as if it were one of the many wonders of the world that gets into the Guiness Book of World Records.

Charles Darwin, in trying to understand laughter and its place in evolution, argued that our response to tickling is not very different from our response to a joke. One must be in a 'light' state of mind to be able to respond to both with laughter, he insisted; and, of course, that surprise was necessary to produce laughter. Did he mean to imply that only those with 'minds' were capable of laughter? We know that many animals are ticklish. But trees?

As scientific research reveals a new history of emotions in plant studies, and as books such as this one continue to explore the human-plant interface, I wait to hear the 'laughter of the trees'.

Sumana Roy
27th August, 2023

Acknowledgements

The timing of this project on Human-plant coincided with the growing Russian invasion of Ukraine, particularly amid the news of Ukrainian troops regaining control of the Chernobyl nuclear power plant after Russian forces had occupied it for over five weeks or so. The news of Russian troops allegedly suffering from a high-frequency radiational level more than 30 years after the dreadful accident makes one wonder how the same place, known for its hostile radioactivity to humans, can come to represent Europe's third-largest nature reserve with its large silver birches, vines and meandering bushes completely overtaking the areas around nuclear catastrophe. The project takes its calling from many such instances of vegetal resiliencies to human-induced forces that have, through its planetary engineering, irrevocably changed the trajectory of this planet and pressed the panic button of apocalypse in the fragile corridors of our planet.

This volume seeks to capture the human-plant dialogues vectored around the debates of the Anthropocene. The dialogues as such have been aptly enlivened by eleven sparklingly nuanced chapters, all of which are singularly distinct in their tonal range as well as in stylistic variations, that open up on fresh new phyto-centric imaginaries and insights as responses to the Anthropocene. We wish to acknowledge all our contributors here, deeply and sincerely. Heartfelt thanks go out to Sumana Roy for adding a 'touching note' to this volume; the resonance of her thinking on plant-life, throughout this book, will always be something more than we can possibly hope to acknowledge here. We are forever indebted to Ranjan Ghosh for horizoning a 'world' where one always feels the throb and excitement of entering newer thoughts, cognitions and discoveries. Finally, we greatly appreciate the generous and unflagging support of everyone associated with the project at Brill: our editor for this series, the outstanding Michael Marder, Erika Mandarino, the Acquisitions Editor, Helena Schöb, the Associate Editor, and Kayla Griffin, the Production Editor; all of them deserve a special mention for their continuous support throughout. Last but not the least, we stand obliged to the anonymous reviewer(s) for the most insightful comments and suggestions for the book. The hard work put together by everyone on board this project is but the one side of coin: we earnestly look forward to our readers to bring it to a full circle.

Ratul Nandi, Jagannath Basu and Jayjit Sarkar
30 March 2024

Notes on Contributors

Carlos Fernando Teixeira Alves

After completing PhD in History, Carlos Fernando Teixeira Alves has devoted his career to investigating the history of science, knowledge, and medicine, utilizing innovative methodologies and approaches to shed new light on these fields. His doctoral research resulted in the development of an analysis scheme that allows for the comparison of scientific and pedagogical aspects in educational institutions, which has proven useful in understanding the evolution of educational systems over time. One area of particular interest to him is the main actors of pedagogical and scientific reforms, and how they have served as a link between governmental authorities and decision-making bodies of the university. He is particularly interested in examining the role of professors in promoting change during periods of reform, as they are often the ones who have the most direct contact with students and are therefore best positioned to effect change. In addition, his research highlights the close relationship between the university and society, as many reforms were initiated due to public health and economic development concerns. His work provides valuable insights into how the role of the university has evolved over the centuries, and the various reforms that have occurred to shape it into what it is today. Finally, his research also explores the importance of nature and natural resources in the pre-Industrial Revolution period, and the identification of an "official" understanding of the role of nature during this time.

Carolin Becklas

Carolin Becklas is a PhD candidate at the University of Oldenburg, Germany. Her broader research interests include (green) game studies, ecocriticism, scientific illustration and animation, and critical future studies. Her dissertation is a multidisciplinary project that asks how and when climate change is literally brought into (digital) play, how global warming is simulated and represented and what ecological futures are imagined in the games. In addition to her studies, she teaches at the Jade University (Wilhelmshaven, Germany) on the topic of sustainability in digital games.

Juan Manuel Ávila Conejo

Juan Manuel Ávila Conejo is a PhD candidate at New York University. He holds an M.A. in Comparative Literature from the University of Texas at Austin, an M.Phil from New York University, and a degree in Architecture. His research interests include: Latin America & the Caribbean, Francophone literature, Russia literature, the Global South, marxist and dialectical criticism, environmental humanities, ecocriticism, critical theory, film and visual art, and architecture and urban studies.

Sindhura Dutta
Sindhura Dutta is a PhD candidate at Bankura University, India. She also teaches at Asleha Girls' College. She has done her Masters from University of North Bengal, and has completed M.Phil from Vidyasagar University. She is an academic editor at *New Literaria: An International Journal of Interdisciplinary Studies in Humanities*. Her research interests are Environmental humanities, Postmodernism, Northeast Indian Literature, Dystopian Literature, Science Fiction, Victorian Literature and Art movements in literature.

Nesrin Eruysal
Nesrin Eruysal received her doctoral degree in American Culture and Literature from Hacettepe University with a dissertation entitled "Red Lions and White Lilies: Mysterium Coniunctionis (Spiritual Matrimony) in H.D. and Robert Duncan's Selective Poems in the Light of Jungian Psychology". Specializing in Jungian literary criticism, she is primarily interested in Jungian Analytical Psychology, American Poetry, and Translation. She has served as editor of Joseph Conrad and His Works (Book of Proceedings of the 10th METU British Novelists Seminar). She has published two volumes of the Turkish version of Conversation Poetry Quarterly, a digital poetry journal born in 2007 in Canterbury, UK through the dialogue of a small group of poets. Her translations of contemporary Turkish poetry have appeared in many journals, including The Wolf (UK), Söyleşi Üç Aylik Şiir Dergisi / Conversation Poetry Quarterly (Turkey), Silk Road a literary crossroads (USA), Visions International (USA), Tower Journal (USA), The Literary Review: An International Journal of Contemporary Writing (USA), Qarrtsiluni online literary magazine, Poetas del Mundo, Turkish Poetry Today (UK) and Mediterranean Poetry an odyssey through the Mediterranean world. Throughout her teaching career, she has taught courses at the undergraduate and graduate level in translation and literary theories. She has also worked as a researcher at the Poetry/Rare Books Collection, State University of New York, Buffalo, and a visiting scholar at the Department of English Literature at Lehigh University (PA, USA). Her most recent position is Acting Chair, Assistant Professor and Translator at OSTIM Technical University. Her email id is: nesrineruysal@gmail.com.

Andrea Färber

Andrea Färber is a PhD candidate at the American Department of English and American Studies at the University of Graz, Austria. She works as a research assistant at the Centre for Intermediality Studies in Graz (CIMIG) and teaches literary studies to undergraduate students. Her research interests particularly focus on climate change literature and the representation of and interaction with the more-than-human in literary texts. She is the deputy chairwoman of TINT, an ESL Literary Journal and Association.

James Jack

James Jack is an American Asian artist who weaves stories of place, spirit and resilience. Art projects have been exhibited at documenta fifteen, Setouchi Triennale, Busan Biennale Sea Art Festival, Centre for Contemporary Art Singapore, Honolulu Museum of Art, Donkey Mill Art Center and Tokyo Metropolitan Museum of Art. Published articles include "Perepetetic in the Pandemic" *art journal open*, "Dirt Stories" *ANTENNAE*, "Spirits of Tsureshima" *Shima* as well as writings in books such as *American Art in Asia, Place-Labor-Capital* and *Mono-ha: Requiem for the Sun*. Jack holds a PhD in art practice (Tokyo University of the Arts), was a Crown Prince Akihito scholar (University of Hawai'i at Mānoa), Postdoctoral fellow at Social Art Lab (Kyushu University), Georgette Chen fellow (Yale-NUS College) and is Associate Professor of Intermedia Art and Science (Waseda University). www.jamesjack.org

Perkus Chun-kit Leung

Perkus Chun-kit Leung has completed the M.Phil. programme at the Department of English Language and Literature of Hong Kong Shue Yan University. He also received an M.A. in Interdisciplinary Cultural Studies from the same department. His M.Phil. thesis examines the intersection between William Wordsworth's poetry and the emergent affect theory through a Deleuzian lens. His research interests include ecocriticisms, affect theory, and the intersection between Daoist and Deleuzian philosophies.

Alexandra M. Peck

Dr. Alexandra M. Peck is the inaugural Audain Chair in Historical Indigenous Art at the University of British Columbia's Art History, Visual Art, & Theory Department. As an anthropologist and material culture specialist, Peck examines pre-colonial and colonial-era Native art of the Pacific Northwest. Her research explores how art reflects cultural change and social interactions in

Indigenous societies in Washington and British Columbia. She received her PhD. in Anthropology from Brown University, where she studied with Drs. Patricia Rubertone and Robert Preucel. Previously, she served as Visiting Scholar of Indigenous Studies through the Mellon Environmental Stewardship, Place, & Community Initiative at the University of Minnesota-Twin Cities. Her work has featured in the *Journal of American History*, *Texte Zur Kunst*, *Journal of Northwest Anthropology*, *Manual: A Journal about Art & Its Making*, and *Open Rivers: Rethinking Water, Place, & Community*, and has been funded by the Wenner-Gren Foundation for Anthropological Research and the Reed Foundation. In addition, she serves as co-editor of *Archaeology in Washington*.

Jacob Price

Jacob Price is a native to the Pacific Northwest in the United States, who combines his love of the outdoors with his fascination with literature. Currently a Visiting Professor at Brigham Young University, his research includes examining the relationship between humans and nonhumans across a variety of art, including interactive media such as video games.

Clara Soudan

Clara Soudan received her PhD in environmental philosophy from the University of Edinburgh. Her doctoral dissertation analyses the gnostic motive of a dualistic alienation from the world and unfolds a philosophy of entangled overflowing—embracing both the openness of a world of becoming and our radical entanglement with more-than-humans. Drawn to questions of intimacy, liminality and apophasis in our relationship to more-than-human life, her research delves into environmental philosophy's fertile intersections with critical theory, plant studies, gender studies, ecocriticism, process theology and quantum physics.

Figures and Tables

Figures

Tables

Introduction

In his book, *The Botany of Desire* (2001), Michael Pollan invites his readers to participate with him in an exhilarating thought-experiment: he wished his readers to imagine and explore the very rationale behind the vegetal landscape (primarily garden plants), amid which they mostly find themselves, from a perspective which is typically its 'own', which is to say a point of view from within its leaves, roots and stems. Such a 'plant-eye view' of the world, according to Pollan, triggers an extraordinary epistemic shift in which an apparently human endeavour like gardening would reveal itself to be not typically a human invention but as an event 'manipulated' by a plant in order to spread its genes by way of luring the human as well as insect pollinators through its wealth of colour, fragrance, texture and shape. Though it cannot deny its affinity with some form of '*strategic anthropomorphism*',[1] Pollan's agentialzing of the plant's point of view nonetheless does best to interrogate and expose the veiled anthropocentrism that underpins human thought since millennia: a belief that uncritically allows humans to inscribe their point of view as 'sovereign' over all kinds of plant bodies. What this phyto-centric perspective does is to open the question of the point of view to a radical uncertainty: Do the humans really domesticate the plants or vice versa? Is gardening an event instituted by plants or humans or both? Is this at all about the question of who domesticated whom? Thinking along such lines can, all of a sudden, make us perceive a phenomenon like agriculture not as an exclusive human invention, nor typically as result intended and contrived by plants, but as a co-evolutionary technology in which both human and plant share a kind of an intimate '*being-with*' or togetherness: an entanglement so profound yet so commonplace that human beings can only ignore at their own peril.

Of course, from a co-evolutionary perspective, human and plant life have always been entangled. The entanglement as such almost stretches back to the earliest of times when the emergence of Cyanobacteria, roughly some 3,500 million years ago, had set the stage for the very condition of human life along with animals. Indeed, for the last twenty millennia, humans and plants have co-evolved in such diverse but intimate ways that the history of one would

1 A term introduced into theoretical parlance by Jane Bennet. Widely reproved as a mark of human imperiousness, Bennet finds in the 'anthropomorphism' a strategy which can help us understand the materiality around us as "agentic" and lead to an energetic politics of our ecological wholeness. See *Vibrant Matter: A Political Ecology of Things*, Duke University Press, 2010.

 | DOI:10.1163/9789004721777_002

be unthinkable without the history of the other. The narrative of the emergence of human civilization on earth bespeaks a silent and often unrecognized reliance of human labour upon the vegetal world—a form of primordially shared 'we-ness' that cannot precisely be made available in comprehensive epistemic terms. This deep unvoiced or unarticulated co-dependency between plant and human world, as Michel Serres insists, is the sign of an immortal pact that natural world signs with the human world—a contract that overwrites human 'social contract'—which underlines the idea that vegetal world is not the 'other' of human world, but its symbiont. Indeed, if we allow the history of the earth to be told, a history which the humans invariably dub as their own, then the vegetal should aptly be seen as a joint author of this historical narrative, collaborating with humans on equal terms.

Yet, today at a time when our forests are burning, coral reefs are fast disappearing, biodiversity declining, the glaciers melting—at a moment of history when human appropriation of plant life has almost reached a planetary magnitude, humanity seems all the more intent on investigating the origin and ramifications of this entangled human-plant evolutionary history. It is impossible to overlook the paradox underlying the situation here. Our current (re) turn to the vegetal paradoxically coincides with a moment in history that has now witnessed a catastrophic global extinction of countless plant species at almost an unprecedented scale. If the findings of a very recent survey are to be considered, then nearly 650 plant species across the globe have been driven to extinction over a span of the last 250 years with approximately one million plant species standing on the cusp of extinction, an event that can easily have a cascading impact on the extinction of all other species thriving on this planet, including humans. The cause of such colossal upheaval is attributable mostly to the human-induced destructive practices such as deforestation, inordinate changes in the use of lands etc. It is very much likely, given the speed and enormity with which the anthropogenic operations have sent myriads of our vegetal symbionts to their geological trash bins of history, that next 100 years will see the end of many of our terrestrial plants that have been here on this plant for last 500 million years and have silently cradled and nourished human civilization for countless centuries. This substantially explains humanity's current obsession with the word 'Green' and why it has become the single most important expression of our time, enjoying unfettered dissemination into varied contexts—socio-political, economic, aesthetic—and creating a new mode of ethical urgency.[2] Bruised with a sense of deep self-reproval and faced with the threat of plausible extinction, humans, thus, have eventually

2 See Andrew Dobson, *Green Political Thought: An Introduction*, (Routledge, 1990).

learnt to make an obsession out of their worst fear, the fear of being extinct. Staying green now has become a survivalist urgency, a holy injunction which, as politics of optimism make us believe, might return us to our lost paradise, to a time in human history when human-plant relationship was supposedly 'natural'. However, in a world ravaged by irreversible shift in climate, loss in biodiversity, incremental rise in carbon-emission rates, uncontrollable proliferation in the genetic modification of plants among others, the tropes of nature and natural can hardly execute a province of unspoiled leafy expanse that once served as a dense and immovable background of human activity. As the Anthropocene assigns normativity and planetarity to the humans, as the frontier between the natural and the human comes to be questioned, as our language of conservation begins increasingly to echo as reparative measures; the poetics of human-plant intimacy today shows signs of having entered a new epoch, writ large with paradoxes yet relentlessly seeking redemption and re-enchantment through a re-awakened phyto-consciousness.

The present volume titled *Human-Plant Entanglement: Thinking with Plants in the Anthropocene* aspires to bring the two urgent paradigms, Plant Studies and the Anthropocene into a close intellectual embrace. By bringing the Critical Plant Studies into dialogue with the Anthropocene, this volume seeks to explore and extend the breadth and the scope of 'crosspollination' between the two paradigms with the intention to effectuate a range of creative-critical perspectives on the idea of human-plant entanglement. It attempts, on one hand, to get to the 'root' of at once contradictory and comforting implications of the current turn to the study of plants amid global environmental crisis, extinction of earth's biomass, genetically engineered crops, and rapacious agronomics. But it also, on the other hand, seeks to probe how many of the momentous epistemic assumptions made available by the lexicon of the Anthropocene (and its other cognate terms like Capitalocene, Plantationocene, and Chthulucene etc.) inflect the burgeoning field of Critical Plant Studies. It explores the broad breadth and depth of such heterogeneous ideas through a register that is truly cross-disciplinary, working at the intersections between the differential domains of aesthetics, philosophy, politics and science. But mostly, it endeavours to substantiate and 'cultivate' an ardent belief that attentiveness to plant life can allow our habitual anthropocentric and instrumental assumptions to be invaded by a unique *phytocentric* proposition that promises new ethical imaginary for a human-plant relationship and inspires new ways of conceiving human-plant 'growing-with'.[3]

3 A wordplay on Jean-Luc Nancy's concept of *'Being-with'*.

In seeking to bring 'plant-thinking' at the thick our critical consciousness, in looking to foreground the vegetal currency back—from its shadowy ready-to-hand obscurity—at the centre of all modes of human cogitation in a way that mediates nearly all forms of planetary concerns : from the crisis of climate change to the question of sustainability of the planet itself, the *Critical Plant Studies* has truly made it possible to address and think through the implications of human-plant entanglements in an array of uncommon contexts. It has enabled us to focus exclusively on the dynamics of human-plant relationship away from the holistic, broad-brush, and conceptually loose denominations like 'the environment' in which it is embedded. While the term 'environment', as a composite of distinct set of human and nonhuman *actants* (Latour), primarily treats the vegetal as static green backdrop for the fast-paced human activity, critical plant studies shift its attention exclusively towards the specificities and nuances of vegetal life which has the potency to subvert and transform the monolithic holism of much contemporary environmental criticism or Ecocriticism. It brings our focus back solely to the question of the individual plant-life and to the implication such a distinct 'way of being' carries for our thought. Such concepts of vegetality, in turn, challenges our conceptual rigidities of vegetal life and reanimates the ethico-political stakes of our otherwise moribund and routine epistemic behaviours.

Our relationship with the vegetal life has hardly ever moved beyond the calculus of exploitation. As a species, humans have always, as Michel Serres thinks, parasited the vegetal world, gorged on its manifold and unparsimonious offerings without scarcely stopping to appreciate the plant for the singularity, abundance, and multiplicity with which it confronts us. In its avowed urge to think *with* the plant, away from the totalising and reductive pull of metaphysical thinking, this volume seeks to reach back to the singular plant (not to terms like "Nature" which tends to subsume the singularity of plant life under a regimented and amorphous green principle) to the throbs of its roots, to the restive energy of its cotyledon and to the busy culinary verve of its leaf. The very structure of a plant with its "rooted unfreedom", "non-conscious intentionality" and its "non-teleological growth"[4] signals the possibility of a radically transformative politics that can wage a war against the totalizing ontologies of dualism and lead towards new post-metaphysical imaginary characterized by a unique vegetal turn. What makes a plant so unique is its irreducibly polymorphic identity, i.e., its ability to lend itself to grafting (both physical and conceptual) whereby it multiplies its personality even as it remains firmly

4 Marder, *Plant-Thinking*, 2013.

rooted in its place. Johann Wolfgang von Goethe, the renowned German poet and polymath, in his botanical poem *Metamorphosis of Plants* (1970) touches upon this astonishing capacity of a single cell of a plant to project itself through multiple forms: petals, stamens, styles and finally the fruit. The self-repeating and morphological mechanism of Goethe's archetypal plant *Urpflanze*, which he goes on to find amid sea of other plant species, is what he identifies as the ultimate 'Plant-Ocean' (*Pflanzen-Ozean*), conceived as the very cradle of all life on this planet.

Interestingly, Goethe's work apprises the 'leaf' (*blatt*) as the ultimate example of his Primal Plant: a singular unit of a plant that can also be plural. If 'everything is leaf' to begin with, as Goethe believed, then one experiences it at the microscopic level of mosses and lichens as well as at the large-scale level of forests: from the small-scale arthropods to planetary bipeds like humans, it can sustain and nourish all forms of life. It is this ontological trajectory of a plant life wherein it acts as simultaneously *singular-plural*, which allows it freely to move up and down the scale of organic life, that it helps to open a vegetal counterpoint to the Anthropocene in respect of the range of scale.

Of all aspects one usually associates with the Anthropocene, it is the scalar dimension that comes across as the most intimidating. The mammoth scale of the geological shifts induced by human actions on earth began with the Neolithic Revolution roughly 8000 years ago, an event that had been the game-changer. The agriculture has given Neolithic people a *techne* to scale up their productivity of crops, their underlying motto being "more existing is always better than any quality of existing".[5] Having set the stage for commercialized and incipient capitalized agriculture, the Neolithic age had set the ball rolling for the next big stage in the history of human dominance: the Industrial Revolution in the eighteenth century. The shift from agrarian and sedentary forms of economy to wholesale machine-based industrial economy had changed the anatomy of the entire planet through unlocking a form of energy-source that altered human technologies pretty much forever: fossil fuel. The uncurbed mining of hydrocarbon deposits (formed over millions of years) for almost over two centuries, as much as it led to an unprecedented scale and speed of human technology, also precipitated a moment in which the humans have become a new geological stratum, leaving its indelible carbon-footprint on earth forever, one that will *post-date* human beings in the distant future. Thus the story of finite humans becoming geophysical force of inhuman scale (in the manner of a tectonic plates or a volcano) is also the story of how they

5 Morton, *Humankind*, 50.

have done terrible things to earth in order to ramp up their commercial profits, scale up industrial productivity to satisfy their unending demands for wealth accumulation: an apprehension in which the Anthropocene is best understood as Capitalocene and where anthropogenic climate change is seen inescapably as *capitalogenic* climate change. However, what is very obvious is that, even when one desists from vilifying the capitalist economy for its destructive ecocidal trajectory and relays the responsibility onto certain forms socialist governments, human planetarity has reached a momentum which is impossible to override i.e., humans lack any flexile mechanism to minimize or downscale the extent of damage it caused to the planet.

In contrast to turbulent and stupefying scale of human colonization of the planet, the vegetal colonization (which occurred approximately over a span of 350000 years) of the planet had been by rather quiet, unostentatious, and largely compassionate to the vast territorial expanse over which it spread: it was by far the most genial, benign, and selfless form of colonialism one could ever think of. Monica Gagliano, Patricia Viera and John C. Ryan in cutting-edge work on plants, *The Language of Plants: Science, Philosophy*, put it quite succinctly:

> It was the silent yet relentless colonization of terrestrial environments by the earliest land plants that transformed the global landscape and radically altered the geochemical cycles of the planet. This resulted in lowered concentrations of atmospheric carbon dioxide and thus set the scene for the emergence of terrestrial animals about 350 million years ago.[6]

In comparison, the Anthropocene scale is customized towards exponential growth in the mass productivity, exploitation, use and management of terrestrial resources following a stringent unilateral trajectory that only knows how to 'scale up' everything with or without the need to do so. The ceaseless and often careless exploitation of natural resources which led to a cataclysmic modification of this planet has almost made the planet unsustainable; in its relentless forward-moving acceleration, fed and fuelled by a misplaced arrogance in what Timothy Morton calls "the anthropocentric copyright control,"[7] human planetary consequences have reached a stage that can neither be controlled nor be appropriately measured. In contrast, the plant scale operates not only on an equally powerful or even broader planetary scale but it seems

6 Gagliano, Ryan and Viera, *The Language of Plants*, 2017, vii.
7 Morton, *Dark Ecology*, 80.

to operate on a completely different ontological register, one which is compellingly ethical in its scope and ramification. Unlike the Anthropocene scale, it can vertically shift up and down ladder of life, (very much like Goethe's transcendental plant) exuding a remarkable morphogenetic plasticity: it manifests itself easily through the enormity of large trees that assemble to make up the great arboreal forest as well as through the diminutive scale of mosses, lichens which spread their lush and carpeted layer underneath our feet, ever conspicuous yet always overlooked. In sliding back and forth between the different scales—between the molar and the molecular—it assumes a stewardship to earth in a way that perpetually lets life on the planet *be*. Heather I. Sullivan is right to point out the life-affirming vitality of the vegetal scale which is markedly distinguished from the devastative planet-altering scale of the Anthropocene. Her observations are very incisive:

> It is this vision of scale that I juxtapose to the global vision of the Anthropocene as human up-scaling through industrialization (that relies on down-scaling livability and the value of life, especially of the non-human). Attaining a geological scale in the Anthropocene means reaching devastating levels of alterations to our living world. Let us seek here instead to think multidirectionality, non-humanly, to think of/like/with plants.[8]

Turning our attention to vegetal scale demands an upending of our *all too human* understanding and at the same time attuning ourselves to a completely alien account of 'worlding' in which ideas such as temporality, thinking, activity, survival would enunciate something entirely different than the one we are conditioned into. For instance, the idea of time in the Anthropocene is tied to speed; human commitment to live by the ordinance of industrial time, to thrive in a world made increasingly smaller and accessible by the techno-scientific advances have made us understand time only in terms of its 'accelerated' rhythm. But with plants, we are catapulted into a domain of existence where time feels dilated: the operative momentum of vegetal life is set to a sluggish and almost lazy tune which not only pits itself against the 'clock-time' which normativizes human experience of the duration (of time) and sets out to counter the dismissive undertone of some of our vocabularies such as 'lazy' or 'inactive' seem to carry by giving them a positive inflexion for self-justification. Contrary to the Anthropocene scale, scalar dimension of plant life works

8 Sullivan, "Goethe's Leaf", 2021.

by accommodating a 'delay' into its ways of being and doing: Plants help to improve the quality of soil by shedding leaves which assists in aerating and enhancing the soil, but such a 'processual duration' is definitely not anthropogenic. The same goes for the photosynthesizing marine phytoplanktons (serving humans with the most popular form of energy source i.e., 'petroleum') which has metamorphosed into its present hydrocarbon form over a course of millions of years through a geological accident that defies human sense of time. The lackadaisical and languid cadence of plantscapes bear a mute and an enduring testimony to a vegetal takeover of the planet that took place millions of years before the advent of the *Anthropos* and its strident mechanical-time. It is precisely this reason which has led to 'plant blindness' in our modern industrialized culture, since a human eye is not adapted to detecting the slower and stealth mobility of a plant; in the vegetal world, time is oriented towards an unfamiliar temporal-scale: what means an entire year to us (stretching to three sixty-five days) may only be a day (twenty-four hours) to a plant. Thus, it is gratuitous to claim that the motion with which the world unfolds before a human observer will be the same to plants and other organisms.

Like temporality, vegetal life similarly prepares us for a radically defamiliarized sense of words like 'thinking' and 'survival'. The idea of thinking without involving any human agency or any self-regulated neuronal structure may be inconceivable to human minds but certainly not to plants. As Michael Marder in his extraordinary book *Plant-Thinking: A Philosophy of Vegetal Life* reminds us how Western metaphysical thinking has tried to instrumentalize thinking as the sole preserve of a rationally conscious human *cogito*. Marder is quick to point out the plant's way of "non-conscious intentional" living (which is non-dualistic, material, and non-teleological) can open up a singular paradigm of thinking where relation between mind-body, active-passive, conscious-unconscious is not oppositional and where *thinking* and *being* belong together. Such an insight would reveal not only the underlying ethical kernel intrinsic to vegetal life but exhibit, as Marder claims, how such vegetal ethics can lay the foundations for a "post metaphysical" vegetality by challenging and dislodging some of the core anthropocentric assumptions which feed on Western metaphysics. Encouraging us to find *something of the vegetal being in us*,[9] *vegetality* as a concept reminds us that plants are not just blunt materialities extrinsic to our existence but subtend many of their behavioural attributes to the human realm, forcing us to discover the *inner plant* in us—the

9 Marder, *Ibid*, 182.

immanent plant-like quality of human beings that has always been neglected by the dominant metaphysics over a millennium.

Setting ourselves the task to reconnect with our own twigs and tendrils can lead towards astonishing revelations about our collective bio-political lives: the Covid-19 pandemic situation, seen by many as both an event *in* and *of* the Anthropocene, is a clear case in point. The Covid event, the horrid memory of which is still very fresh in our collective consciousness, has added a new epistemic volume to the growing intimacy between the planetary crisis brought about by the Anthropocene and the growing vulnerability and precarity of human lives. The recent studies have shown how global climate change has opened the floodgates for many vector-borne viruses which are likely to cause major threats to the human population as well as to other species in the near future.[10] As Bruno Latour puts forth his worrying hypothesis of how this virus has the potency to act as a "dress rehearsal for the next crisis, the one in which [...] the health crisis prepares, induces, and incites us to prepare for climate change".[11] While the *Coronacene* had the entire world reeling under its tremor, while the sweeping mobility of life got superseded by an unannounced call for a wholesale *stasis* on all walks of life, the deeper entanglement between the human-plant relation began to surface: all of a sudden—as soon as the spirit of living changed from kinesis to stasis, from outdoor to indoor,—the 'inner' vegetality started mirroring the outer. We started realizing how 'rootedness' (to one place) is not unfreedom but can be more life-sustaining and life-ensuring, as the spaces outside became increasingly 'viral'. To our surprise, we also became aware of true vegetal propensities of existence at large when (during the lockdown period we all were stationed indoors) we internalized a type locomotion which was typically plant like : forced to pull out from all actual (physical) modes of communication, we instantly fell upon the *other* means of communication governed by 'virtual' platform through different kinds of (internet-assisted) social or digital mediums that we brought us together like never before; much the way trees branch out and talk to one another through an organized underground mycorrhizal networks of roots called 'Wood Wide Web'. The Covid pandemic as an Anthropocene event, thus, can also be seen as a moment of productive crisis, an event that may also stimulate and provoke us to think more intensively, transitively, and tangentially about the human-plant relation in order to forge newer assemblages.

10 Rocklöv & Dubrow, "Climate change: an enduring challenge", 479.

11 Latour, "Is This a Dress Rehearsal?", 2021.

There is, however, a strong warning given against any attempt to find anything 'progressive'[12] in the Anthropocene other than its legacy of a rapacious "industrial modernity" and its destructive underside: its profound impact on the way it legitimized colonial appropriation of indigenous environmental resources that takes myriad forms. Natasha Myers accurately catalogues such lethal forms environmental enactments stretching over centuries:

> From the clearing of vast amounts of land for sugar, rubber, and cotton plantations made possible through slavery and colonization; to the extraction of vast amounts of petrified photosynthetic life from reservoirs that fed the expansion of the Industrial Revolution and today's petro-capitalisms; to the Green Revolution, which continues to unfurl in a time of late industrialism with the proliferation of chemical herbicides and pesticides in twenty-first century forms of industrial agriculture.[13]

Understanding how colonialism is complicit with the way people in the former European colonies (which amounts to almost 80 percent of the whole world) understood and eventually shaped their relationship with vegetal landscape around them brings the Anthropocene into tangle with the postcolonial or neo-colonial contexts, which chiefly take as its focus the widespread and continued exploitation of both the indigenous population and vegetal resources alike at the hands of Western colonial powers. Indeed, a closer scrutiny of the underlying grammar of the innumerable human-plant interfaces during colonial times reveals the extent to which a number of discursive practices worked in sync with colonial forces to further the interest of the empire: forestry, agriculture and even botanical science—all are tied up with legitimizing the vocabulary of 'profit' into an everyday episteme. The Western colonial administration adopted such a commercialized agricultural strategy that would drain the colonies of their forest resources by clearing the wild vegetation and farming crops that would affect the fertility of native lands yet would ultimately boost their buffer stocks as well as their economy back home. It is important to remember that the ideas of forestry in the former colonies of Europe remain de facto braided with the idea of colonial forestation. Such an

12 For Clive Hamilton, the Anthropocene marks the obvious fall out of European modernity's relentless drive towards self-aggrandizement and self-appropriation of humanity, a move which is deeply tendentious and dangerously ecocidal. See Hamilton's "Defiant Earth: The Fate of Humans in the Anthropocene" (Polity, 2017), 80.

13 Natasha Myers, "Photosynthetic Mattering: Rooting into the Planthroposcene", ed. Line Marie Thorsen (Rønnebæksholm, 2017), 123–129.

operation surprisingly would implicate botanical science within the rhetoric of colonial expansion and consolidation of the empire. Botany—the objective and factual scientific study of plants—in such contexts stands completely integrated in the imperial politics of power as a discourse that primarily originates in great experimental trans-continental sea-expeditions of the British colonial power. The Botanists were, in effect, not solely scientists confined to research-laboratory but actively involved in building a vegetal archive that would promote re-naming a particular form of flora withdrawn from its local origin and helping it transfer across the colonies to be used for largescale plantations. Such a forced plantation, with obvious profit in sight, will invariably invade and supplant the surviving ecoscape native to the colonies and lead ultimately to massive alterations in the environment. The botanical sciences, thus, entered a symbiotic relationship with the forces of colonialism in which it played the role of a benefactor that has precipitated the larger motive of colonial expansion and profit, but also of a beneficiary which has been, in turn, organized by the same expansionist drive it has helped prosper. In sum, the postcolonial or the neo-colonial contexts attempt to counter the universalist assumptions of the Anthropocene as uniformly affecting all human communities everywhere by provincializing the question of victimhood; the idea is to articulate how the global paradigm of the Anthropocene works only by ignoring the fact that perhaps not all human communities on this planet have contributed to the Anthropocene equally and perhaps some vegetal and human landscapes are more exploited and endangered than others (the third world, for instance). However, with China currently emerging as the world's major source of carbon-emissions and further, with many of the former European colonies resorting to neo-colonial modernist drives that would very often conflict with the broader environmental concerns, the idea of sequestering the politics of human-plant (in the Anthropocene) within the geopolitical markers of 'Global South' and 'Global North' does seem to have lost its appeal to some degree and one feels the urge to move towards a new and experimental politics that would step aside the rigid dualistic categorization.

This desire to move beyond the enclaves of dualism towards new forms of human-plant 'being with', a goal which this volume aspires to achieve, requires new forms of imaginaries that can help us not just view but also smell, touch and even taste their manifold interactions (or *intra-actions*) differently across the bounds of disciplinary fetters. It can be as experimental as the modern day Bio artist Joel Ong's speculative research project that seeks to meld the apparently incompatible fields such as the genetic study of plants and literary criticism or it can also take the form of a bizarrely captivating experiment carried out by the Biologist Eduardo Kac at the Weisman Art Museum of Minneapolis in which he has evidently grown a new form of plant named *Edunia* that blends DNA of the petunia flower with that of his own, resulting in an unique molecular signature of a human DNA on the petals of a common flower.[14] The dawn of the Anthropocene seems to have prised open the gateway to new-age scientific epistemes that reorient our approach to plants. The bright side of some of these discoveries has the potential to be the game-changer for our time. Recent findings prove that microalgae can be a great source of sustainable biofuel that can at once solve our energy crisis and help minimize carbon footprint. Another recent development is the flourishing of food-tech companies operating in plant-based meat products. When the climate-based issues are the most salient questions to chime in the corridors of our planet, when even a seemingly unproblematic issue like meat-eating assumes a global proportion (foods extracted from animal products alone account for a significant increase in greenhouse emissions rate) affecting the environment, the choice of a plant-based meat does come across as food of the future. While these scientific endeavours are duly pushing the limits of human experiment with plant sciences at the very moment when we require it the most, one simultaneously needs to attend to the kind of speculative and creative-critical consistency at display in the disciplines of humanities. This volume contends that works of literature and philosophy, written with a remarkable sensitivity to vignettes of plant life, aspire to do much more than just to project the plant-life as a mere poetic or analytical prop for procuring aesthetic and metaphysical knowledge; one needs to explore how they help us discover in the creative flow of life a mutually creative urge of human and plant life, anticipating a strange kinship that goes beyond all dichotomies in thinking.

By urging us to question the *obvious* in the human-plant relationship, this volume endeavours to stimulate us into challenging the potency of our commonplace lexicon we habitually resort to in order to make sense of such a relation. The following extract is from one of Jane Bennet's essay in which she seeks to capture the poetics of proximity between herself and a common invasive plant called *Ailanthus altissima* from a typically interstitial viewpoint that solicits a re-thinking of our existing vocabulary. The conceptual and methodological frameworks articulated within this volume find seamless resonance in Bennet's articulate prose.

14 See Megan Gambino, "The Story of How An Artist Created a Genetic Hybrid of Himself and a Petunia," *Smithsonian Magazine*, February, 22, 2013, https://www.smithsonianmag.com/science-nature/the-story-of-how-an-artist-created-a-genetic-hybrid-of-himself-and-a-petunia-25148544/.

> *Ailanthus* grows itself toward sun, emitting tentacles of indeterminate longing; simultaneously, sun casts down rays upon the warming and expanding concrete and the stretching tree; roadway lends itself as a support for the tree, expressing an inner alliance, an affinity, between its smooth concrete and the surface texture of the bark. Along with *Ailanthus* and roadway on that sunny day, I too was leaning and longing, casting and connecting, radiating and being radiated. Standing under the road, I felt an involuntary surge upward, repeating imprecisely the upward thrust of the tree, my tilted head aligning with the plant's crooked yearning. This was happening as my stance—two feet firmly planted—was resonating with the cool stolidness of concrete. Given that my body is made up of vegetal, stony, mineral, and solar solids, liquids, and gasses, perhaps we can understand the mimetic relays between *Ailanthus*, sun, concrete, and flesh as nonpsychological, reverberating nods. To "nod," "emit," "radiate," "stretch," "lean and long," "cast and connect": are these the verbs appropriate to the action style of an onto-Sympathy?[15]

Works of poetry and literature have always been perceptive to the subtleties and vagaries of vegetal life and served, as Jonathan Bate once said "as imaginary parks in which we may breathe an air that is not toxic and accommodate ourselves to a mode of dwelling that is not alienated."[16] Literary works, by virtue of being inherently non-representational or non-objectifying, can affect a radical *defamiliarization* (to use a word from Russian Formalist critic Roman Jakobson) of our conventional lexicon to create a vocabulary of intimacy between vegetal life and the human world. This inner intimacy between human-plant discerned in the literary works is something that cannot be restricted to any genre, time, or agenda. For instance, the modern-day readers can find a steady line of poets—running from Keats and Walt Whitman to Jonathan Skinner and Denise Levertov to Kamala Das—whose works are interspersed with an acute receptivity to a distorted frontier between human and plant life. What these poets have in common is that they have written about human-plant affinity in the most intimate manner possible without having made such concern a governing political, as it were, agenda of their poetry: the idea of human-plant entanglement is so basic to their works that they did not have to put out any formal manifesto about it. If the vegetal and the human are already parts of a given relationality, is there any need to validate or acknowledge such a relation

15 Bennet, "Of Material Sympathies," 241.
16 Bate, *The Song of the Earth*, 64.

in a sort of a secondary mode of cogitation? Interestingly it is this act of thematizing or *re-cognizing* plant life in literary works that creates a rupture between the two in the first place and induces an objectification or instrumentalization of the vegetal. It is in this sense of bringing a green perspective to poetry that would resist being commodified or hegemonized by the programmatic expressions of eco-consciousness that their works appear distinctly contemporary.

This volume has been meticulously organized into three distinct sections, namely: "Human-Plant and Territoriality", "Human-Plant and the Literary", and "Human-Plant and Transfiguration". Each of these sections holds significant importance in elucidating the multifaceted relationship between humans and plants. Within this structural framework, a total of eleven chapters are distributed. This deliberate arrangement underscores the thematic coherence and scholarly rigour of the volume. The first chapter "Charles Edenshaw's 'Fungus Man' Platters & Tlingit Grave Effigies: Exploring Human-Fungus Relationships through Agarikon Art", authored by Alexandra M. Peck, belongs to the section entitled "Human-Plant and Territoriality". It is a profound inquiry to explore how within Haida and Tlingit societies, life and death are intricately linked to a particular fungus, known as Agarikon (*Laricifomes officinalis*). This chapter explains how this woody conk features prominently in Haida origin stories, where Agarikon is anthropomorphized as the legendary "Fungus Man" character, whose bravery and power are credited with the creation of women. Amongst the Tlingit, Agarikon plays a complementary role and was employed in pre-colonial mortuary settings, where the fungus offered protection for the deceased. Examining 19th century Haida artist Charles Edenshaw's carved argillite platters that depict Fungus Man, as well as ancient Tlingit grave effigies carved from actual Agarikon, this chapter draws upon traditional ecological knowledge, oral history, mycology, and material culture to explore the unique spiritual, symbiotic relationship between fungi and First Nations. As further evidence of its protective qualities, Agarikon possesses properties that fight against tuberculosis and smallpox, both diseases that significantly affected Native nations after contraction from European settlers. Now critically endangered because of logging and timber industries—a visceral reminder of settler colonialism's extractive impact on the Pacific Northwest's lush rainforests—Agarikon's future is uncertain. Nevertheless, the fungus remains spiritually, cosmologically, and culturally significant to the Haida and Tlingit, even as Agarikon becomes increasingly associated with a pseudo-mythical, pre-colonial past.

Just as Agarikon was responsible for creating women, the life-givers and cultural caretakers of Haida society, the fungus was a meaningful Tlingit reminder of death and dying. As we grapple with the modern Anthropocene and accompanying dystopian events (climate change, old growth forest devastation, disease epidemics) that have disproportionately impacted and forever altered Indigenous nations, material reminders of Agarikon offer a new lens through which to view these challenges. Unlike other studies of human-fungi relationships that emphasize solely the medicinal, culinary, or hallucinogenic uses of mushrooms, this case study represents a deeper, more nuanced example of how human and fungus or plant-like entities are intertwined and dependent upon each other. Agarikon reminds us that we cannot reference human origins or even death without acknowledging the agency and influence of local ecology.

The second chapter "How to Coexist: The Teaching of the Plant Kingdom at the Universities of Coimbra and Salamanca (XVIII–XIX)" by Carlos Fernando Teixeira Alves analyzes the construction and evolution of the discipline of Natural History introduced in the Universities of Coimbra and Salamanca. Natural History reflects the official construction of the Human-Nature relationship that influenced the generations that drove the scientific and industrial advances that characterized the Industrial Revolution. Natural History's main goal was to provide extensive knowledge about the various forms of exploitation of natural products, not only of mineral and animal origin, but also of vegetable origin. This relationship (Human-Nature) was intended to facilitate solutions to issues such as poverty, has its systematization in the 16th century, in Italy, and promote the growth and development of human societies. The discipline of Natural History, occupied, until very late (XVIII–XX), a prominent position in the hierarchy of knowledge; dictating what was the technical capacity for the knowledge of Nature and the exploitation of natural products. Based on the relevance of Anthropocene Studies in the reflection of human action on the environment and its impact on ecosystems, we intend to contribute with a historical approach from university teaching between the 18th and 19th century, based on the evolution of Natural History teaching in both Universities. After its introduction in 1772, the chair of natural history eventually fragmented and gave rise to other disciplines such as Botanic and Agriculture. However, the content of this discipline does not change, but expands and starts to enhance new ways of studying plants. But what the case of this universities shows us is that only the study of plants that could be of economic or medical interest was promoted. Through the various statutes this chapter intends to analyze the delicate bond that linked the needs of human beings to natural products which eventually became a decisive component in the learning process.

The third chapter entitled "Tēichigi Colors of Artistic Resistance" by James Jack explores how family lineage of vegetal ancestors are revealed while peeling fresh bark of *Techika* (Japanese Yeddo Hawthorne). Plants *are* ancestors; plants *are* siblings. This chapter seeks to argue that despite American and Japanese soldiers leaving scars across the land, people and trees on Tokunoshima (徳之島, "Island of Benevolence"), working with vegetal friends as "plantcestors" can heal trauma, hurt, and violence. Resistance is rooted in the social fabric of islanders including poet and politician Izumi Hōrō's vision. This non-resistant way of resisting continues today. For example, in 2010 more than half of the island's population assembled in the port to demonstrate against a US base relocation proposal. Archipelagic solidarity was successful in thwarting the relocation plan standing with the broader aim for "no peace without equality" (Chinen 2010).

The traditions of sericulture revealed in dye, cotton and silk crafts over the past five centuries are alive in fibers with an infinite range of earth tones within. This chapter shows that through anthropocentric lenses vegetal relations have been misconstrued away from plants, yet the next generation of craftspeople including Yukihito Kanai continue with philosophies including "dye is attitude, not traditional." Phytocentric worldviews return through art created together with the voices of *techika* and signs of islanders. Weaving the language of islanders in social movements along with folk patterns are combined into an artistic work *Techika Resistance*. It exemplifies that creativity sustains activism in the long term in colorful, engaged and imaginative ways. Seeds nourished in "small" places have large impacts on networks of solidarity. The revival of "traditions" occurs within the complexities of the present while nurturing intimate relations with our vegetal ancestors, botanic siblings and seed offspring.

The fourth chapter "Plant-thinking and Critical Pedagogy in Costa Rica: the Revolutionary Praxis of Carmen Lyra" by Juan Manuel Ávila Conejo talks about Carmen Lyra, who used to be a kindergarten teacher, writer, feminist activist, and founder of the communist party of Costa Rica. Her literary work—profoundly political as well as pedagogical—is informed by the historical juncture of Central America in the 1920's and 1930's: the grand epistemic struggle between the rise of United Fruit Company and American neocolonialism against the growing influence of communism in the region in the wake of the Russian Revolution. Lyra inhabited a world in ruins—in France, as a young woman, as she experienced the human and environmental devastation brought about by World War I; and later encountered the same ecological catastrophe in her home country of Costa Rica, this time a product of economic crisis and plantation agribusiness. In this context, she developed a kind of thinking about pedagogy, literature, and political organizing modeled after weeds, or *matas*: the 'lowest' form of plant life and yet the most adaptable and enduring. This chapter explores her work as an early form of plant-thinking based primarily on her lived experience in Central America as a kindergarten teacher and caretaker of impoverished children, yet also influenced by progressive European and Soviet pedagogical innovations. In order to make this argument, it attempts to read excerpts from her political speeches, pamphlets, and essays in order to weave together a theory of eco-critical, or materialist, pedagogy. This chapter argues that Lyra's critical, plant-based pedagogical practice at the Maternal School constitutes a form of revolutionary praxis that can help us to find a way forward in our age of environmental and political catastrophe.

The chapter "The Root of the Problem: How Trees can be Utilized in Climate Change Literature to Depict the Interconnectedness of Past, Present and Future" by Andrea Färber belongs to the section entitled "Human-Plant and the Literary". It explores how climate change literature grapples with mediating between the issue of future repercussions of global warming and the reader. In doing so, it tries to shift the attention from an anthropocentric worldview towards a new understanding of our planet that aims to displace the culture/nature dichotomy. Climate change literature often focuses on an approach that situates the human within the more-than-human, urging readers to adopt a new perspective on nature that places the human within it rather than keeping it as separate from and superior to it. With climate change being considered a slow violence (Nixon) whose full extent will only become visible over the course of decades if not centuries, the ramifications are still unimaginable to most people and thus the issue is being ignored for the time being. It is therefore of the utmost importance to find suitable ways to help the audience at large understand that these inconceivable threats already affect the present because of deeds done by our ancestors, and that today's actions will have even greater repercussions on the generations yet to come. In order to portray these spans of time that human perceive as vast in a way that allows the reader to understand how much of an impact human interference has on the planet, a literary agent is needed that can connect the present to both the future and the past. Trees offer themselves as the ideal facilitators for this task. Through the analysis of Annie Proulx's novel *Barkskins* and Richard Power's novel *The Overstory*, this chapter aims to achieve two things: First, it explores how trees can be utilized in climate change literature in order to disrupt an anthropocentric worldview and displace the human main characters. Second, it suggests that to make people understand the issue of time in the context of climate change, trees lend themselves as mediators because of their longevity and the impact they have on all forms of life. Through tracing the history of trees and their entanglement with human lives and cultures over centuries, as well as weaving research on trees into the stories that have the potential to permanently influence one's world knowledge, these climate change novels, this chapter argues, attempt to demonstrate both the impact and dependence that humanity has on these more-than-human beings.

The sixth chapter entitled "Multispecies Exchange Rates: Multicultural Plant Entanglement in *The Embrace of the Serpent*" by Jacob Price shows how in contrast to capitalistic political systems that value plants by their economic and human-benefitting properties, the 2015 film *Embrace of the Serpent* positions humanity's dependence on vegetable life as a type of currency for multicultural exchange. In each of the film's two timelines, protagonists Karamakate, Theo, Evan, and Manduca have their own goals and understandings of plant life in the Amazon, but each come to depend on each other's knowledge to fulfil their desires. Their timelines and plant knowledge entangle throughout the film, as each person renegotiates their vision of the Amazon when in conflict with others'. Karamakate, the only character to appear in both timelines of the movie, functions as the nexus of the multicultural exchange between European science, evolving indigenous identities, American extractivism, and resource management during World War II.

This chapter analyses how the ideological, cultural, and plant knowledge in *Embrace of the Serpent* reflect the greater issues and entanglements of the indigenous, scientific, and cultural judgment values of plant life in the Amazon. The competition of these valuations critique and celebrate how globalization has functioned as a catalyst of ecological and ontological change for all cultures that have bought into the world economy through coercion or their own desire. But as these valuations clash, they also create opportunity for people from inside and outside Amazonian cultures to reconcile their differences through their dependency on plants found in the jungle. These negotiations require multicultural exchange rates for each character to understand and help each out. Their conversations and interactions with plant life in the Amazon result in multispecies exchange through shamanism in ways that challenge historically exotifying or othering perceptions of indigenous knowledges and cultures.

Nesrin Eruysal's chapter "Enchanting Encounters with Plants in Reshma Aquil and Sumana Roy's Poems" explores the literary works of Indian poets Reshma Aquil and Sumana Roy, by utilizing the theoretical framework of "phytocriticism" as defined by John Charles Ryan in his book *Plants in Contemporary Poetry: Ecocriticism and the Botanical Imagination*. By drawing on Arvind Krishna Mehrotra's contemplations on the topography of Aquil's childhood home and the surrounding "forest", this chapter argues that Aquil's early interactions with plant life supports Gaston Bachelard's proposition that childhood is vegetal. The first section of this chapter investigates Aquil's poems from *Sleeping Wind, Shadows of Fire*, and *The Unblending*, which illuminate her revelations of unexpected connections between nature and culture and situates them within the contexts of both Emanuele Coccia and Gaston Bachelard's analyses of arborescent life. The second section offers a close reading of Sumana Roy's books, *How I Became a Tree*, and *V.I.P Very Important Plant*, to demonstrate the stages of humans becoming trees, or vice versa. Roy's poem on Bose's old house, which is replete with arboreal imagery, and whose archetypal roots can be traced back to traditional religious and philosophical encounters with plant life in India, is examined in the light of Jagadish Chandra Bose's theory of botanical consciousness.

The next chapter "Flowering-with the Silence of Plant: Affective Attuning to the Vegetal Event in Tang Poetry" by Perkus Chun-kit Leung states the significance of plants in Chinese poetics, be it owing to the preponderance of vegetal imageries, or to the fact that many Chinese poets drew inspiration from plants. In Chinese poetry, particularly the landscape poetry composed in the Tang dynasty, the plant world is a manifestation of the Buddhist teachings of "purifying the heart and illumining one's true essence 淨心明性" and the Daoist ideal of self-so-ness 自然. More specifically in relation to the rethinking of the Anthropocene with plants, vegetal silence in Chinese poetry becomes a germinative site that echoes Marder's notion of a genuine plant-thinking as adumbrated in *Plant-Thinking: A Philosophy of Vegetal Life*: "let[s] plants maintain their otherness" while "respecting the uniqueness of their existence."

The aim of this chapter is to examine the silence of plants in Chinese poetry in search for a lively plant-voice that resists Anthropocenic discourses about anthropocentrism and humanism. To adumbrate the scenario of attuning to the silence of vegetal life, this paper takes Marder's idea of a vegetal event as a conceptual threshold to a genuine perception of the botanical rhythms and agencies through affect. This chapter resorts also to the Buddhist idea of emptiness and the Daoist notion of non-action for proffering a non-Western mode of plant-thinking. Through examining the non-anthropocentric portrayal of plants in selected Tang poems, this chapter argues that an eventful encounter with plants resides in an attunement to the lacunae and silences of plants, wherein the vegetal silence emerges as an inter-rhythm that orchestrates our imbricated rootedness to this kaleidoscopic world of non-human agencies.

The chapter "Vegetal Apophasis: Reclaiming Dark Intimacies and Un/knowing with Plants" by Clara Soudan belongs to the third and final section: "Human-Plant and Transfiguration". This chapter attempts to show how dissolving boundaries between cultural and natural, inside and outside, mind and matter, the ecological apocalypse manifests our radical entanglement within more-than-human lives. World-composers between the earth and the sky, intimate strangers breathing us into being, plants embody the collapse of a modern, dualistic order of nature. Today inextricably imprinted onto the earth's crust, our words and thoughts condition deeply entangled ways of knowing and unfurl into multi-species processes of world-composing. Attending to these more-than-human intimacies unearths porous, mutual inhabitations which overflow dominant regimes of thought. Such intimate entanglements do not exhaust the strangeness, opacity or distance of that which inhabits with and within us, but rather demands that we espouse a radical intimacy with radical strangers.

This chapter thus, seeks to ask: How does the awareness of our terrestrial entanglement with more-than-human lives profoundly alter the way we think and know? How may we render ourselves sensitive to the cryptic ways in which more-than-humans address us? How could we cede room for what withdraws itself from our perception and understanding? Drawing upon the work of Michael Marder, Natasha Myers, Catherine Keller, Karen Barad, David Abram and Timothy Morton, it weaves an ecological thought committed to that which overflows the epistemological regime of nature. This chapter believes that vegetal lives summon us to hold a space for other ways of thinking, knowing and storying. The apophatic thought unfolded in this chapter suggests that plants escape our attempts to seize them, compelling a renewed attention to their own modes of sensing and knowing. The prism of an apophatic ecological thought thus offers an alternative epistemological framework which, undermining the solipsist pitfalls of anthropocentrism, opens ways of knowing in the intimate, obscure presence of more-than-humans. Cultivating an awareness to what eludes present regimes of knowledge, vegetal apophasis delves into the ways in which plants, subverting dominant narratives and structures of thought, summon us to un/think and un/know. What might a reclaimed intimacy with the obscurity of vegetal life disclose that present regimes of thought eclipse? Teaching us to regenerate the process of philosophical thought, the earthly darkness of plants invites us to reclaim an aesthetical, epistemological and political sensitivity to the presence and wisdom of that which escapes or is repressed from our attention. Embracing their withdrawn dimension, apophatic languages of unsaying might allow us to clear a space for a renewed encounter with plants, in an intimacy respectful of their obscurity.

The chapter titled "To be a Pixel Flower: Player-Plant Entanglement in Video Games" by Carolin Becklas attempts to show how Video games offer imaginative, interactive, and exploratory spaces in which the player can experience new perspectives on the interplay of human and more-than-human world. In these digital worlds, survival is usually impossible without plants. They serve as important resources, such as raw materials for buildings or essential food, indispensable for crafting medicine for the avatar. In "sustainability games," they are also CO_2 reservoirs that absorb harmful greenhouse gases and thus ensure the well-being of the virtual ecosystem. Plants can even be found in the procedural structure of the game itself: as so-called "tech trees" they serve as a visualization of a series of technologies which can be build and advanced as the player levels up. Therefore, what Emily Dickinson wrote so eloquently in one of her poems holds true in games: "To Be a Flower, is profound / Responsibility—." This chapter thus explores what kind of "responsibilities" do plants assume in video games by means of a few selected examples and thereby mapping these crucial player-plant entanglements which eventually shape the way we make sense of our "real-world" in an interwoven ecosystem.

The final chapter "Theatrics of the West, and Geopolitical Economy in the Plantationocene: Opium Poppy and its Nation-based Phenomenological Understanding" by Sindhura Dutta attempts to make an Economic study of global trade of narcotics, specifically Opium, that shows the scales tipped when it comes to consumption of opium in its different forms. China was the 19th century's leading consumer of opium, whereas in the 21st century, America is the leading consumer. Wars have been fought over opium, its cultivation, illegal trafficking, and all of them sum up to only one desire, the desire to dominate globally. We live in the era of the Plantationocene, as Donna Haraway describes in her book *Staying with Trouble*. This compels us to investigate our relationship with the poppy plant, which was first cultivated in the Mediterranean region. Both Opium and the Plantationocene have one thing in common, that is Britain. It thoroughly commodified Opium through its trade with China and progressed with the epoch of the Plantationocene. This led to the Anglo-Sino War, and impoverished Indian farmers because of its coercive agricultural policies. America today is the world's largest consumer of opium, which ironically failed to regulate its production in Afghanistan despite several wars; indicating such crackdowns are only theatrical attempts of drug regulation. Plants have evolved for their progenitive purpose and we have coevolved with them, the common purpose being survival. The latex or opiate tapped from pods to prepare substances too, is the plant's defence mechanism to protect itself from pests. Likewise, in the Anthropocene, nations have colonized the opium poppy to progress economically, whose ultimate goal is to progress its nation-based race. Therefore, the human-plant relationship and the historical study of opium poppy reposits our relationship with them. The aim of this chapter therefore, is twofold. First is to conduct a phenomenological study of poppy plants in terms of experience of different nations. For example, opium poppy cultivation, consumption and its trafficking are experienced separately by India, China and Britain respectively and also in literary fictions of the country such as *Cherry*, *Ibis Triology*, *The Opium Prince* and *Confessions*. Likewise, how America and Afghanistan experience or share a relationship with this plant will be different from each other. Such a study will focus on different experiences of the same plant by different nations based on the types of experience it has had. Secondly, a Plantationocene critique of Opium Poppy Cultivation and its consequences in Global South countries which is likely to reveal the subjugation of farmers and environmental degeneration in the era of cash crop plant cultivation, thereby showing how colonial rulers have used this plant to subjugate the people of its colonies.

These chapters admirably explore how in our drama of everyday life the vegetal and the human are thrown up against each other, perhaps more deeply than we can ever anticipate. This volume is a kind of a work that straddles genres—memoir, botanical science, literary genres, philosophy, anthropology, history, video games and what not—to *grow* a narrative of intimacy between the vegetal and the human, offering a profound creative-critical mediation on the ways two are always deeply intertwined, an experience not exhausted either by the vocabulary of bureaucratic green or by our metaphysical urge to see them split into an overly fetishized binary of 'self and other'. This book certainly offers (what philosopher Michael Marder would call) a true *post-metaphysical vegetal ethics* that comes to think of both the vegetal and human as proximate intra-corporealities whose existence very often gets tangled up in each other's folds and becomes a story of indistinct togetherness. Indeed, this is exactly the kind of ethical articulation we need to foster in our times of anthropocenic emergency.

Bibliography

Bate, Jonathan. *The Song of the Earth*. Picador, 2001.

Bennet, Jane. *Vibrant Matter: A Political Ecology of Things*. Duke University Press, 2010.

Bennet, Jane. "Of Material sympathies, Paracelsus, and Whitman," in *Material Ecocriticism*, edited by Serenella Iovino and Serpil Oppermann, Indiana University Press, 2014.

Gagliano, Monica, Ryan, John C. & Vieira, Patricia. *The Language of Plants: Science, Philosophy, Literature*. University of Minnesota Press, 2017.

Gambino, Megan. "The Story of How an Artist Created a Genetic Hybrid of Himself and a Petunia." *Smithsonian Magazine*, February, 22, 2013. https://www.smithsonianmag.com/science-nature/the-story-of-how-an-artist-created-a-genetic-hybrid-of-himself-and-a-petunia-25148544/.

Hamilton, Clive. *Defiant Earth: The Fate of Humans in the Anthropocene*. Polity, 2017.

Latour, Bruno. "Is This a Dress Rehearsal?". *Critical Inquiry*, Vol. 47, No. S2 (Winter 2021): https://doi.org/10.1086/711428.

Marder, Michael. *Plant-Thinking: A Philosophy of Vegetal Life*. Columbia University Press, 2013.

Morton, Timothy. *Dark Ecology: For a Logic of Future Coexistence*. Columbia University Press, 2016.

Morton, Timothy. *Humankind: Solidarity with Nonhuman People*. Verso, 2017.

Myers, Natasha. Photosynthetic Mattering: Rooting into the Planthroposcene, edited by Line Marie Thorsen, Rønnebæksholm, 2017.

Rocklöv, Joacim & Dubrow, Robert. "Climate change: an enduring challenge for vector-borne disease prevention and control". *Nature Immunology*, 21 (29 April 2020): 479–483. https://doi.org/10.1038/s41590-020-0648-y.

Roy, Sumana. *How I Became a Tree*. Aleph Book Company, 2017.

Sullivan, Heather I. "Goethe's 'Leaf' and Scales of the Anthropocene: The Vegetal versus the Geological." *The Philosophical Life of Plants*, Fall, 2021. https://www.plantphilosophy.org.uk/j-w-goethe-philosopher-botanist/goethes-leaf-and-scales-of-the-anthropocene-the-vegetal-versus-the-geological/.

PART 1

Human-Plant and Territoriality

CHAPTER 1

Charles Edenshaw's "Fungus Man" Platters and Tlingit Grave Effigies: Exploring Human-Fungus Relationships through Agarikon Art

Alexandra M. Peck

1 The Anthropocene and … Agarikon?

A seemingly inescapable phenomenon, the Anthropocene is an increasingly ubiquitous term used to refer to our current era—marked by a rise in use of plastics, significant global climate change, heavy metals, species extinction, and nuclear waste.[1] In short, an era of significant human influence resulting in disastrous consequences for humans and non-humans alike. Although some scholars have tried to pinpoint the exact origins of the Anthropocene, others find this a futile exercise and instead propose that the Anthropocene is better understood as a culmination of various events and attitudes.[2] One such event is the non-Indigenous settling of the Americas and the devastating impacts of settler colonialism to Indigenous environments.[3] Although pre-colonial Native civilizations manipulated and developed their local landscapes, these practices were significantly less destructive than those introduced by settler colonists to North America.[4] Following the lead of Native scholars such as Zoe Todd (Metis) and Kyle Whyte (Potawatomi), some propose that we re-label the Anthropocene as the "Kleptocene" and define the term by settler colonial theft of land and resources, early disease epidemics that impacted millions of Indigenous individuals, and other violent forces that led to our current ecological emergency.[5] Likewise, feminist theorists highlight patriarchy and capitalism as significant contributors to the Anthropocene, prompting us to interrogate

1 Zalasiewicz et al, "When Did the Anthropocene Begin?"; Steffen et al, "The Trajectory of the Anthropocene."

2 Crutzen & Stoermer, "The 'Anthropocene'"; Zalasiewicz et al, "Are We Now Living in the Anthropocene?"; Doughty et al, "Biophyscial Feedbacks."

3 Bauer et al, "Anthropocene"; Gibbard et al, "The Anthropocene as an Event."

4 Lewis & Maslin, *The Human Planet*; De Loughrey, *Allegories of the Anthropocene*; Keeler, "Before Colonization."

5 Whyte, "Is it Colonial Déjà Vu?"; Davis & Todd, "On the Importance of a Date"; Lightfoot et al, "European Colonialism"; Baldwin & Erickson, "Introduction"; Guernsey, "The Infrastructures."

 | DOI:10.1163/9789004721777_003

how male dominance, Manifest Destiny, and misogyny influenced the settler colonial treatment of Indigenous communities, cultures, and landscapes.[6]

Currently endangered, Agarikon (*Laricifomes officinalis*) is a jeopardized fungus native to the increasingly rare old growth forests of North America, Morocco, Europe and Asia (fig. 1.1). The large, woody conk perches on larches, fir, spruce, hemlock, and pine trees, where it initially presents as a humble, albeit majestic, fungus.[7] However, Agarikon offers more than meets the eye. As a threatened polypore with significant cultural ties to the Northwest Coast, specifically the Haida and Tlingit Nations of Haida Gwaii (Canada) and Alaska's panhandle, Agarikon serves as a prime example of how humans and their environment are delicately intertwined. Featured prominently in epic Haida origin stories, where Agarikon is anthropomorphized as the legendary "Fungus Man" character, the conk's bravery and power are credited with the creation of women during a pre-gendered time. Amongst the neighboring Tlingit, Agarikon plays a complementary function and was employed in pre-colonial mortuary settings, where the carved fungus offered protection and peace for the deceased. Further enhancing its cultural significance, Agarikon contains medicinal properties that link the conk to disease epidemics introduced by settler colonialism.

Examining 19th century Haida artist Da.a Xiigang Idinsaw's (Charles Edenshaw's) argillite stone platters that depict Fungus Man, as well as Tlingit grave effigies carved from actual pieces of Agarikon, this chapter draws upon traditional ecological knowledge, oral history, mycology, and material culture to explore the unique spiritual, symbiotic relationship between fungi and First Nations. In doing so, Agarikon emerges as an ideal lens for examining the role of Indigenous cosmology in relation to local ecosystems. Today, Anthropocentric consequences inform many Indigenous livelihoods, who bear the brunt of rising sea levels, melting glaciers and ruthless forest fires, increase in invasive species, and loss of biodiversity. How then, do we grapple with and attempt to mitigate the repercussions of the Anthropocene? Indigenous scholars such as Winona LaDuke (White Earth Ojibwe) and Daniel Wildcat (Yuchi-Muscogee) argue that possessing spiritual or cosmological connections to the natural world is necessary for the success of modern environmental movements.[8] If our mythological and religious worlds are intimately tied to the natural environment through legendary sagas, ethical systems, and omnipotent non-human

6 Haraway, "Anthropocene"; Moore, *Anthropocene or Capitalocene?*

7 Blanchette et al, "Extraordinary Fungal Masks."

8 LaDuke, *Recovering the Sacred*; Wildcat, *Red Alert!*

FIGURE 1.1 Agarikon

entities, they argue, then we are more motivated to protect the sacred. Doing so avoids what Robin Wall Kimmerer (Potawatomi) describes as "species loneliness," the estrangement or isolation of humans from other species.[9] After all, we rarely destroy what we deem sacred.

2 Overlooking the Importance of Mushrooms for Northwest Coast Native Nations

Assessing Indigenous Northwest Coast attitudes towards mushrooms is necessary before delving deeper into Agarikon's significance. Depicting fungi, or even plants, is an unusual choice within the Native Northwest art arena, making the embrace of a conk unique for many reasons. Instead, animals, such as Raven, Wolf, and Eagle, are common designs for Haida and Tlingit artists. The Northwest Coast's limited representation of mushrooms may be attributed to "mycophobic tendencies" or myco-antipathy, reinforced by certain Northwest Indigenous culinary traditions that only utilized mushrooms in times of scarcity or emergency.[10] Perhaps this reluctance alludes to the high skill level needed to locate, correctly identify, and prepare mushrooms, as the Pacific Northwest contains a variety of toxic species.[11] Moreover, mushrooms are regarded "apprehensively" in various Northwest Coast Indigenous languages, which associate fungi with death, blindness, corpses, harm to children, bad luck, severe weather, or ghosts.[12] On the other hand, some languages rarely differentiate between various types of mushrooms, instead referring to all fungi in general terms, implying that mushrooms as a whole are rather unremarkable. For instance, the Haida language classifies all mushrooms using non-specific monikers such as k'ak'w dajáangaa ("bogeywoman" or "witch" hats), kagan dajáangaa (mouse's hat), or st'aw dajáangaa (Saw-whet owl's hat).[13]

Despite prevalent notions of mushrooms as taboo or not particularly interesting, these explanations fail to acknowledge mycological significance in household settings. Evidence suggests that many prized mushroom varieties,

9 Kimmerer, *Braiding Sweetgrass*; Govindrajan, *Animal Intimacies*; Gagne, *Caring for Glaciers*.

10 Compton, "'It Pulls Everything to You,'" 6; Turner, "Economic Importance," *Food Plants, Ancient Pathways*; Galloway, *Upper Sto:lo Ethnobotany*; Turner & Kuhnlein, "Camas & Riceroot," *Traditional Plant Foods*; Kari, *Tanaina Plantlore*; Turner et al, *Ethnobotany*.

11 Turner & Kuhnlein, *Traditional Plant Foods*.

12 Ibid; Compton, "'It Pulls Everything to You'"; Compton et al, "Puffballs from the Past"; Nicholson, "Some Spiritual Uses," 27; Turner, *Ancient Pathways*, 313; Turner & Cuerrier, "'Frog's Umbrella.'"

13 Lachler, "Dictionary of Alaskan Haida."

including chanterelles, morels, oyster mushrooms, American matsutake, are highly valued and consumed by Northwest Coast First Nations.[14] Other fungi are utilized to create blankets or multi-colored pigments for cosmetics and paint.[15] These examples illustrate the importance of fungi in Indigenous settings, countering the assumption that mushrooms held only negative connotations along the Northwest Coast.

Based on the plethora of evidence that suggests otherwise, characterizing Northwest Coast attitudes towards mushrooms as hesitant or indifferent is erroneous. In addition, linking mushrooms to a presumed fear of toxic or hallucinogenic properties does not accurately align with the millennia of scientific knowledge that Native communities possess. Yet, emphasizing only the culinary or practical uses of mushrooms risks portraying Native Northwest relationships with fungi as solely sustenance or utility-based, thus valuing the secular over the profane. A lack of scholarly attention paid to deeper, more complex uses of fungi points toward ethnographic ignorance or bias, as plant pathologist Robert Blanchette argues: "The lack of familiarity of early ethnologists with fungi and the primary focus of their studies on higher plants are likely reasons that information about fungi is lacking in most older works."[16] Downplaying the role of fungus serves to strengthen the Anthropocentric notion that nature is inevitably growing less and less relevant to Indigenous societies. Despite the myriad integration of mushrooms in Northwest Coast settings—and despite stewarding a famously damp, rainforest ecosystem—many scholars perpetuate the myth that fungi played a lesser role amongst Northwest Coast First Nations, inadvertently legitimizing the near extinction of certain species (including Agarikon). Instead, as this chapter proposes, fungi possessed very different cultural significance than many are familiar with today.

3 Raven and Fungus Man Create Men and Women

To better understand the complex role of fungi in Haida society, we must start at the beginning. Haida origin stories feature primordial Yaał (Raven), wandering aimlessly after the Great Flood, waiting for the waters to subside.[17]

14 Turner & Bell, "The Ethnobotany of the Coast Salish"; Turner, *Food Plants, Ancient Pathways*; Galloway, *Upper Sto:lo Ethnobotany*; Turner et al "The Cottonwood Mushroom," *Ethnobotany*; Turner & Kuhnlein, *Traditional Plant Foods*.

15 Teit, *The Shuswap, Mythology*; Jenness, *The Saanitch Indians*; Turner & Bell, "The Ethnobotany of the Coast Salish"; Turner & Kuhnlein, *Traditional Plant Foods*; Turner, *Plant Technology, Ancient Pathways*.

16 Blanchette, "Haploporus Odorus," 233.

17 Dawson, *Report*; Chittenden, *Official Report*.

Lonely, he nurtures and marries a cockle shell, only to hear faint cries coming from within.[18] Curious, he cracks it open to find six infants emerging from the cockle's watery center.[19] Unexpectedly bestowed with the gift of parenthood after impregnating this cockle, Raven fathers the first humans! The newborns mature rapidly, growing into adults who beseech Raven to find them suitable mates.

According to artist Charles Edenshaw (1839–1920), who relayed this story to anthropologist Franz Boas in 1897, the special shell responsible for carrying the first people to term was a cockle—the humble clam's distant cousin.[20] So central to Haida cosmology is this origin story that renowned Haida artist Yaahl Sgwansung, Bill Reid (1920–1998), carved a larger than life version, with countless other artists re-creating the legend in painted and print form (fig. 1.2). Yet, a more diligent analysis of this story uncovers a critical detail: the first humans created by Raven and the cockle were all men, leaving the creation of women shrouded in mystery.[21] Indeed, upon closer inspection, Reid's monumental sculpture depicts equally colossal male genitalia.

Somewhat surprisingly, the Haida associate female origins with conks. Known also as polypores, conks are woody fungi that grow on trees and are characterized by their flat, white underbelly and brown, crusted growth rings. Colloquially, Northwest Coast Indigenous languages compare conks with bread or biscuits, owing to their resemblance to these baked goods. According to Nuu-chah-nulth oral history, when Captain James Cook arrived in March 1778 with pilot biscuits or hardtack, Indigenous speakers likened the food items to conks.[22] Similarly, in Tlingit, Haida, and Tsimshian, conks were labeled "tree biscuit," "wood-pilot biscuit," "pilot biscuit's grandmother," "wood cracker," or, simply, "bread."[23] The Haida, however, also refer to conks as "tree scallops," alluding to the polypores' shell-white exterior and ribbed edges.[24] This latter association of conks with shellfish is particularly notable because it points towards a pre-colonial term, whereas the bready connotation denotes a more recent trend. In addition, the resemblance between scallop shells and cockle shells raises the possibility of a gendered parallel between the tale of Raven

18 Dawson, *Report*.

19 Hoover, "Charles Edenshaw"; Browne, *Entering Time*.

20 Boas, *Primitive Art*; Browne, *Entering Time*.

21 Chittenden, *Official Report*.

22 Marsden, *Na Amwaaltga Ts'msiyeen*.

23 Blanchette et al, "19th Century Shaman Grave Guardians," 119; Lachler, "Dictionary of Alaskan Haida"; Rogers, "Medicinal Mycology," 30–31; Turner, *Ancient Pathways*, 216.

24 Turner, *Ancient Pathways*.

FIGURE 1.2 Raven & the First Men by Bill Reid (1978)

D. GORDON E. ROBERTSON, CC BY-SA 3.0 <HTTPS://CREATIVECOMMONS.ORG/LICENSES/BY-SA/3.0>, VIA WIKIMEDIA COMMONS

copulating with a cockle to create men, and an epic Haida saga where Raven and a conk are dispatched to create women.

Reid told how Raven grew bored with the first men and liberated them from their cockle shell home, prompting him to search for a new source of entertainment.[25] Skaay, or John Sky (1827–1905), a Haida informant for anthropologist John Swanton, elaborated on this story in 1900. He recollected how Raven spotted a small island where brilliant red chitons (beetle-shaped mollusks) grew atop the rocks, bearing an uncanny resemblance to a certain female body part (figs. 1.3–1.4). Surrounded by rough waters and a powerful spiritual forcefield, the island and its chitons repelled Raven's canoe. Undeterred, Raven enlisted the help of various other birds to accompany him on his chiton-gathering expedition. He placed Junco in his canoe, and, spear in hand, attempted to pierce the chitons. As they approached, Junco "'became deranged,'" causing the frenzied pair to turn around and return defeated.[26] Next, Raven convinced his comrade, Kl'aay Kl'aay (Steller's Jay), to join his crew. When they neared the island, Steller's Jay "'flapped his wings around,'"

25 Bringhurst & Reid, *The Raven Steals the Light*.

26 Rogers, "Medicinal Mycology," 31.

FIGURE 1.3 Chiton's exterior shell

KIRT L. ONTHANK, CC BY-SA 3.0 <HTTPS://CREATIVECOMMONS.ORG/LICENSES/BY-SA/3.0>, VIA WIKIMEDIA COMMONS

FIGURE 1.4 Chiton's interior body

JERRY KIRKHART FROM LOS OSOS, CALIF., CC BY 2.0 <HTTPS://CREATIVECOMMONS.ORG/LICENSES/BY/2.0>, VIA WIKIMEDIA COMMONS

creating an uncontrollable stir.[27] Raven then "tried all creatures in vain," dissatisfied that all of his friends "became different" when confronted with the chitons' inexplicable aura.[28]

Raven persisted with his chaotic endeavor, struggling to secure a sensible, level-headed friend to help him reach the chitons. Dejected, he resorted to desperation. With a sturdy twig, he "drew a design on a bracket fungus," or, "'wood biscuit.'"[29] His doodles formed a lively face, but he was anxious to resume his voyage, carving out only crude arms and legs for the conk. Like a magic spell, Raven animated the conk by commanding the fungus to "bestir yourself" and "keep your eyes open."[30] Wedging his new creation into the stern of the canoe, Raven gendered the polypore, naming him Galaga Snaanga, or, Fungus Man.[31] Raven instructed Fungus Man to steady the canoe as they approached shore, no matter how turbulent the current.[32] Initially compliant, Fungus Man began to shake his head and moan as they reached the island, regretting this involuntary adventure. Nevertheless, with Fungus Man's aid, Raven retrieved two chitons. Although impacted by the chitons' spiritual energy, Fungus Man proved sturdy enough to serve as Raven's trusty companion. As they canoed back to the mainland, Raven threw the chitons at the male humans he had abandoned. The mollusks adhered to their skin, transforming into female genitalia. Satisfied with himself and his odd choice of a polypore helmsman, Raven and Fungus Man had created women.

4 Gendered Accounts of a Sexual Story

Male accounts claim that women did not exist prior to Raven and Fungus Man's escapade, instead asserting that men transformed into women after being assigned new sex organs. Upon noticing their feminine resemblance, the "men" became "'ashamed and ran'" to the longhouses.[33] This narrative associates femininity with domesticity and female bodies with shame, in contrast to Haida women's reputation as mobile, public, authoritative figures.[34] Raven assumes a God-like figure in the story, mimicking the biblical "Adam's rib" scenario, complete with naked embarrassment and exile from an Edenic

27 Turner, *Ancient Pathways*, 263.
28 Swanton, *Haida Texts*, 126.
29 Hoover, "Charles Edenshaw," 64; Enrico, *Skidegate Haida Myths*, 63.
30 Swanton, *Haida Texts*, 126; Enrico, *Skidegate Haida Myths*, 63.
31 Ransom, "Art as Negotiation"; Enrico, *Skidegate Haida Myths*.
32 Swanton, *Haida Texts*, 126; Enrico, *Skidegate Haida Myths*, 63.
33 Ransom, "Art as Negotiation," 41.
34 Collison, "Those With Clever Hands."

realm. These similarities led some to question if the narrative was sanitized to appease a Christian audience unaccustomed to sexually explicit scenes and gender-bending descriptions.[35] Charles Edenshaw's daughter, Jadał q'egəngá (Florence Edenshaw Davidson (1896–1993)), offered an equivalent insight that reinterprets the story's final scene as one of male fear: "'Once women change their life, they [men] are scared of them.'"[36]

Likewise, masculine versions timidly describe female genitalia, using euphemisms such as chitons.[37] Some renditions avoid sex entirely, telling how Raven collected the first men in his beak and dropped them near boulders covered in "large but soft-lipped mollusks," where they spontaneously created Haida babies.[38] Other renditions portray Raven and the first men as "rescuing" the women they found "clinging helplessly to the rocks," without any mention of shellfish.[39] In contrast to the accounts provided by male narrators, Haida women's versions of the story reveal previously omitted details, including references to female sexuality and agency. Gid7ahl-gudsllaay Lalaxaaygans (Terri-Lynn Williams-Davidson), General Counsel to the Haida Nation and Haida performance artist, recalls that in the deep past, "Women did not have tsaw [genitals]."[40] Despite this, powerful ancestresses, such as Cloud Woman and Foam Woman, "conceived and gave birth to children without men and without sex."[41] Although men and women both existed during this time, they were apparently unsexed and instead differentiated in other ways.

Williams-Davidson's version demonstrates female bodily autonomy that is absent from other retellings by revealing how women "asked Raven to help them" retrieve female genitals from Tsaw Gwaayaay (Vagina or Vulva Island).[42] Raven first attempted to retrieve the female genitalia solo, but he xaawlagihl (became sweet, a Haida euphemism for orgasm). The power of the island "was too strong," causing him to collapse and tumble out of the canoe.[43] Junco and Steller's Jay succumbed, as well, prompting Raven to enlist Fungus Man as a steersman. Despite being affected by the island's power, "he didn't fall overboard because he was wedged into the canoe [...] he swayed side to side, rhythmically uttering, 'unh, unh, unh, unh,'" as he paddled Raven to the

35 Williams-Davidson, "How Raven Gave Females"; Browne, *Entering Time*.
36 Blackman, *During My Time*, 29.
37 Rogers, "Medicinal Mycology," 31.
38 Bringhurst & Reid, *The Raven Steals the Light*, 35–36.
39 Chittenden, *Official Report*, 20.
40 Williams-Davidson, "How Raven Gave Females," 61.
41 Augaitis et al, *Raven Travelling*; Browne, *Entering Time*, 33.
42 Williams-Davidson, "How Raven Gave Females," 61.
43 Ibid.

forbidden island.[44] After acquiring two sets of genitals—"one large and one small"—from the rocks, Raven and Fungus Man returned to find Eagle Woman and Raven Woman eagerly awaiting their new additions.[45] Each woman was given a set of genitalia, endowing them with the ability to mother the two moieties (Eagle and Raven), from which all Haida individuals descend. In comparison to vague masculine accounts, this version provides a confident feminist analysis where women enlist Raven to physically distinguish themselves from men at the expense of his own safety.

Like the modern female rendition, 19th century Haida storytellers emphasized that Raven sought genitalia for his wife and sister, giving the larger chitons to his spouse, Eagle Woman.[46] This "bigger is better" refrain is predictable when one considers the male narrator and Raven's self-serving persona. His sister, Siwaas (Raven Woman), laments her more diminutive parts, but Raven reassures her that her smaller parts "will be safe [from me]!"[47] Here, Raven refers to exogamy, reinforcing Haida taboos against incest or having intercourse with someone from the same moiety. Yet, rarely do we hear what Raven does next, likely because the account is sexually violent and obscene. Known to be a shape-shifter who is frequently consumed by his own desires, Raven tricks his sister into perching on a mushroom in the forest. "As one might expect," the mushroom is not a mushroom at all, but rather, "Raven's member sticking up into the air. Unable to resist, he has buried himself under the leaves."[48] Devastatingly, Raven betrays his sibling, luring her into a false sense of security and breaking his promise that she "will be safe" on account of her smaller sex organs.[49] Although jarring to modern ears, this seldom-told tale of incestuous rape makes the listener deliberately uncomfortable. The story cautions men to reject Raven's example, reinforcing Haida attitudes towards the proper treatment of family members and women.

5 Recovering Fungus Man

This scene is the last time that we hear any mycological mention in Haida legends, with many storytellers omitting Fungus Man from the iconic chiton story altogether. By concealing sexual and bodily themes deemed inappropriate

44 Ibid.
45 Ibid.
46 Swanton, *Haida Texts*.
47 Ibid, 126.
48 Browne, *Entering Time*, 106.
49 Swanton, *Haida Texts*, 126; Enrico *Skidegate Haida Myths*.

for a 19th century social milieu, sanitized versions expunge Fungus Man and decenter women from the script. This whitewashing of Haida history is linked to settler colonialism, in which severing Indigenous ties, especially those of women, to nature was paramount to the "taming" and re-claiming of North America.[50] As a consequence, or perhaps goal, of the Anthropocene, nature becomes further separated from culture, conveniently disentangling humans and plants from each other. After all, if humans are thought to have no connection with local ecosystems, then the environmental degradation of said ecosystems is presumably irrelevant to human communities. As Fungus Man's role in Haida origins suggests, plant-like creatures are an integral part of Haida society, and their erasure from contemporary written and spoken records hints at something more sinister.

To rectify this, we must turn our attention to material and artistic matters—namely, argillite platters carved by Haida artists and Tlingit grave figures fashioned out of conks, both of which rely upon fungus to reveal complex human-mycological relationships on the Northwest Coast. These objects communicate what might have otherwise been lost to time or as a result of literary censorship and language loss. However, interpreting historical Indigenous art, particularly that of the Northwest Coast, requires culturally-specific knowledge to comprehend abstract motifs and covert meanings. Objects serve as mnemonic devices, containing stories and memories that may not be immediately apparent to modern viewers, regardless of cultural background. Without Indigenous insights and historical context, we risk only appreciating the artform for its aesthetic value and neglect how art illuminates unique cultural attitudes towards local ecology.

6 Charles Edenshaw: Carving through Crisis

Today, the oldest and best-known depictions of Fungus Man reside at the Seattle Art Museum, Chicago's Field Museum of Natural History, and the National Museum of Ireland in Dublin. Charles Edenshaw, the adept Haida carver, crafted these argillite renditions, known as the "Fungus Man platters" (figs. 1.5–1.7). Born in 1839 to a high-ranking Haida family at the village of Skidegate, Edenshaw entered the world during a shifting era (fig. 1.8). Tense interactions with fur traders and explorers began in the late 1700s, but

50 Peck, "Mariners, Makers, Matriarchs."

FIGURE 1.5
Charles Edenshaw, shown carving
Charles Edenshaw
WIKIMEDIA COMMONS, CANADA /UNITED STATES PUBLIC DOMAIN, HTTPS://COMMONS.WIKIMEDIA.ORG/WIKI/FILE:CHARLES_EDENSHAW.JPG

FIGURE 1.6 Charles Edenshaw's Seattle "Fungus Man" Platter
Platter (qwa. a. qiihlaa), 91.1.127, Charles Edenshaw, First Nations, Haida, 1839–1920, ca. 1885, Argillite, 2 ¼ × 12 15/16 in. (5.7 × 32.9 cm), Gift of John H. Hauberg
PHOTOGRAPH BY PAUL MACAPIA, COURTESY OF THE SEATTLE ART MUSEUM

FIGURE 1.7 Charles Edenshaw's Dublin "Fungus Man" Platter
AE/1894.704
PHOTOGRAPH COURTESY OF THE NATIONAL MUSEUM OF IRELAND. © NATIONAL MUSEUM OF IRELAND

FIGURE 1.8 Haida Village at Skidegate, 1878
CANADIAN MUSEUM OF HISTORY, CC BY-SA 4.0 <HTTPS://CREATIVECOMMONS.ORG/LICENSES/BY-SA/4.0>, VIA WIKIMEDIA COMMONS

Haida economies, health, and land claims remained relatively intact until the mid-19th century. Edenshaw grew up surrounded by lavish potlatches, material wealth, and fluent use of the Haida language. He was bestowed with numerous Haida names reflective of his privileged upbringing, tattooed with familial crest designs, and later became a hereditary chief. Regarded as the "best carver and painter … amongst the Haida" by anthropologists and museum collectors, Edenshaw's detailed argillite carvings, bentwood boxes, and totem poles were informed by his intimate cultural knowledge.[51]

In 1862, Edenshaw's world was drastically altered by a smallpox outbreak. The death of valued elders and children dealt a severe blow to his community. In 1885, the Canadian government imposed a legal prohibition on potlatch gatherings traditionally used to redistribute wealth, celebrate marriages, and fortify diplomatic relations. Haida individuals who defied the ban faced imprisonment, while Christian missionaries burned wooden masks and regalia, and federal agents removed totem poles from Haida villages. Likely resulting from Canada's assimilative efforts, it was also in 1885 that Edenshaw and his wife, K'woiyeng (Isabella Edenshaw) (1842–1926), were baptized and remarried in the Anglican Church. The couple chose the royal names Charles and Isabella for themselves, befitting their high status in Haida society.[52] However, Isabella initially objected to adopting Edenshaw as her surname, reinterpreting her name change as a disavowal of Haida law, which banned marriage between members of the same moiety (interpreted by Isabella as the same last name).[53] Citing the Raven tale that warned against incest, she expressed dismay at the Anglican missionaries' ignorance in promoting immoral Haida customs while aiming to "save" Native individuals from their supposed "savage state."

Despite these profound societal changes, Charles Edenshaw managed to preserve important elements of Haida life. To maintain his culture and earn a living, he provided anthropologists with hundreds of stories and commentary on Northwest Coast museum collections.[54] By the early 1900s, at Franz Boas' request, Charles Edenshaw created commissioned works for the American Museum of Natural History in New York. Whereas previous Haida artists made items for community members or trade with other Indigenous nations, a shift occurred during Edenshaw's lifetime in which "'most of his work would have been for the [European market]—anthropologists were going crazy for

51 Boas, *Primitive Art*, 212; Cole, *Captured Heritage*.

52 Blackman, *During My Time*.

53 Davidson, "Reclaiming Haida Culture," 97.

54 Ransom, "Art as Negotiation."

everything,'" according to Jaalen Edenshaw.[55] Art became a full-time occupation for Edenshaw, while his wife, a celebrated spruce root weaver, worked outside of the home at canneries.[56] Combined with the impact of smallpox, missionization, and general cultural decline, Edenshaw's newfound market offered hope, as well as the "'freedom to be an artist in the Western sense of the word.'"[57]

Prior to European arrival, Haida individuals numbered over 20,000, spread across more than 50 villages.[58] Gidansda Guujaaw (Gary Edenshaw), seventh president of the Haida Nation (1999–2012), states that, "'Our people went from 50,000 to 500 in three years. The smallpox came from a passenger [left behind by a European ship].'"[59] In 1900, almost 40 years after the 1862 smallpox epidemic, ethnographers remarked that the Haida population had "'shrunk to not over 700, of whom 300 are here [in Skidegate].'"[60] Missionaries were "'instrumental in having all of the old [long]houses destroyed,'" they noted, with Haida culture now lacking "'everything in short that makes life worth living.'"[61] Remarkably, Charles Edenshaw survived this time period, continuing to carve Haida stories and preserve Haida customs. As his great grandson Guud San Glans (Robert Davidson) explains, "'Everything that was banned, he was doing. Smallpox was killing everybody off. You're losing *villages*, and for him to carry our art into today ... there's just a passion!'"[62]

Of Edenshaw's 11 children, four daughters reached adulthood, with his only surviving son drowning in 1896. Threatened by vicious smallpox outbreaks, Edenshaw was quarantined in the late 1800s.[63] It was during this mournful and uncertain time when he carved all three Fungus Man platters, although it is unclear if the pieces were commissioned or if one platter inspired the others. While the platters share similarities, they are not identical, raising the question of whether Edenshaw was experimenting with different imagery or carving different versions to cater to varying consumer tastes. Despite this ambiguity, Edenshaw's choice of Fungus Man and Raven as his subjects is clear, suggesting that the ribald story was a timely one. But, why portray such a risque tale during a time of cultural duress, with missionaries and government officials making

55 Collison, "Creating for Culture," 25.
56 Blackman, *During My Time*.
57 Collison, "Creating for Culture," 25.
58 Bringhurst, *A Story as Sharp as a Knife*.
59 Stamets, "Antiviral Activity," 2.
60 Bringhurst, *A Story as Sharp as a Knife*, 75.
61 Ibid.
62 Collison, "Creating for Culture," 25–26.
63 Browne, *Entering Time*.

every effort to quash Haida culture? More obvious answers include Edenshaw striving to protect Haida art and cultural values during the onslaught of settler colonialism, or boosting Haida morale during a bleak period. Yet, neither of these explanations fully satisfies. Instead, the answer lies in those elusive and oft-misunderstood chitons. By covertly communicating a heroic Haida story about sex and creation to his peers, Edenshaw emphasized the importance of hope—perhaps encouraging reproduction and vitality when his community's population hit an all-time low.[64]

7 Deciphering the Fungus Man Platters

Of Edenshaw's three platters, the Seattle and Dublin platter are most alike, containing many details and combining intricate Haida formline with European-style facial features. This results in a visual story that may be challenging for those unversed in Haida cosmology. Even noted anthropologist Boas criticized Edenshaw's hybrid designs, decrying the lack of a "pure" Haida style and lamenting that Edenshaw's depictions were rarely realistic or natural looking.[65] Given the enigmatic nature of the Fungus Man platters, it appears that Edenshaw was trying to resist Boas' critique by emphasizing distinctly Haida elements and while simultaneously challenging the Northwest Coast Indigenous art canon. Thus, resistance to settler colonial powers defines Edenshaw's Fungus Man platters—whether by transmitting censored stories, advocating childbirth in the face of death, or utilizing cryptic imagery.

As the story tells, Raven hastily carved a face on a conk, springing Fungus Man to life. Men had only recently been delivered from a cockle shell, meaning that Fungus Man represents the first carving. His rudimentary appearance implies this, hinting that Raven's artistic talents were not yet up to par with Edenshaw's fine-tuned skill level. Alluding to his bready, polypore origins, Fungus Man's Haida name translates literally to "Ugly Biscuits," poking fun at his strange appearance. With large eyes communicating fear or shock, an open mouth voicing a silent scream, arms thrown into the air, and an ill-defined body crouched in the canoe, Fungus Man is a sight to behold. He is a paradox: his pudgy, ambiguous figure seems more fitting for a toddler than a creature with "Man" in his title, yet he somehow bares a full set of teeth and numerous growth rings around his face. This discrepancy indicates his orgasmic state, in which he is transforming from a raw or prepubescent fungus into a

---

64 Ibid.

65 Boas, *Primitive Art*, 158.

fully-formed adult as he nears the female chitons. The conk-like rings framing Fungus Man's face double as a "halo of sparks flying from … the supernatural genitalia," according to Robert Davidson.[66] Despite overcoming his orgasmic state to aid Raven, Fungus Man struggles against the chitons' power, nearly rendering him useless.

As a bracket fungus "accustomed to growing quietly on a tree in the middle of the forest," Fungus Man is taken aback by this sexual encounter.[67] However, like the many puzzles that this story contains, "gender is not [explicitly] marked in Haida pronouns" and is instead marked in other, more contextual ways.[68] This means that an interpreter would have prescribed a gender to the conk, which is precisely what happened: Fungus Man is referred to as "it" until he begins paddling and encounters the feminine energy, when references to Fungus Man suddenly shift to "he."[69] Perhaps disoriented by his new masculine pronouns, Fungus Man has lost his canoe paddle, missing from the platters.

Edenshaw animated the canoe's motion, with Raven and Fungus Man depicted in a pseudo-Cubist style. From one angle, we see the two characters from the front, but a second glance suggests a profile view. Raven perches on the edge of the canoe, wearing a coat of feathers. As a transformer figure with a human-like nose and a long, bird-like beak, his wings are worn "at will, like a garment."[70] His humanoid hands grasp a pointed spear as he approaches the chitons, a bit unsure as he looks back at Fungus Man. The Dublin platter reveals his ovoid-shaped biceps, whereas the Seattle platter shows faint streaks of face paint. Both platters illustrate Raven with long, sometimes braided hair. Like Fungus Man, Raven's portrayal is quite androgynous. The bird is known to transition between male and female forms in other Edenshaw creations, including an articulated mask where Raven appears as a bird on the exterior and a woman on the interior, or a carved argillite chest where Raven resembles a weeping woman cradling a deceased child (fig. 1.9).[71] If Raven's inner or true self is female, then how could Raven "be anything but sympathetic to the pleas" of the primordial women, who tasked him with retrieving their womanly parts?[72]

A large hat shaped like a killer whale dorsal fin protects Raven's luscious locks, referencing a time when "Raven pecked himself out of the body of a whale

66 Williams-Davidson, "How Raven Gave Females," 61.
67 Browne, *Entering Time*, 32.
68 Welsch, "Haida Pronouns," 8; Swanton, *Haida Texts*.
69 Bringhurst, *A Story as Sharp as a Knife*; Browne, *Entering Time*.
70 Dawson, *Report*, 149.
71 Swanton, *Haida Texts*; Wright, *Northern Haida Master Carvers*.
72 Browne, *Entering Time*, 42–43.

FIGURE 1.9 Charles Edenshaw's raven transformation mask

through the end of its dorsal fin."[73] According to Haida belief, the dorsal fin also functions as the world's axis, connecting sea to sky.[74] Knowing this, Raven anticipated traveling through different realms to complete his task. Perhaps he settled on Fungus Man as his final companion because conks reach for the heavens while growing on tree trunks, yet remain firmly embedded within the tree's bark, anchoring them to the earth. Through the melding of elemental realms such as water (canoe), sky (Raven), and earth (Fungus Man), women were created, suggesting that Edenshaw was trying to convey that melding male with female would ensure the continuation of Haida culture and families.[75]

Another realm exists below the water's surface—a darker, more mysterious one. As a "vessel made for crossing boundaries," the canoe is surrounded by an imposing figure lurking underwater.[76] The Dublin and Seattle platters depict this entity as a grub-like creature with an ominous grin—much like a cartoonish chiton. It paces underneath the canoe, waiting to capsize the unsuspecting duo. Robert Davidson posits that the figure represents tsaw sgaanagwaay (the spirit of the chitons, or, women's genitalia).[77]

73 MacDonald, *Haida Art*, 8.
74 Duffek & McLennan, *The Transforming Image.*
75 Browne, *Entering Time.*
76 Ibid, 43.
77 Williams-Davidson, "How Raven Gave Females."

Lastly, images are hidden within the canoe itself. Motifs are shown on the vessel, mirroring painted designs that would adorn a life-sized Haida canoe. Kaagan Jaad (Mouse Woman) is portrayed three times on the Seattle platter's canoe, corresponding to the canoe's three panels.[78] With curly parted hair and big eyes, Mouse Woman was reduced to her current size by Foam Woman.[79] The mighty mouse accompanies those who are traveling to unknown realms or embarking upon long journeys, making her an appropriate design for the canoe. She is a well-known character originating from a story where she helped recover a chief's beautiful daughter, who went missing after refusing multiple suitors.[80] As the story tells, Mouse Woman used her magic sewing awl to cut through the water, with the tool towing the search party's canoes to find the missing woman. On the Seattle and Dublin platters, we see a long, thin implement—an awl—almost unnoticeable beneath the canoe's prow. This design is not mirrored on the opposite side of the platters, and plays no other function except to allude to Mouse Woman. She is a trusty guide, and, like all elements of Edenshaw's platters, it is no coincidence that he included mouse's likeness. According to the Haida, she is the mother of Raven and lives beneath red and white spotted Amanita mushrooms. Appearing below Fungus Man's canoe on the argillite platter, Mouse Woman mimics a mouse seeking shelter beneath a toadstool.

8 The Protective and Medicinal Significance of Agarikon

Edenshaw's platters highlight Fungus Man, honoring the modest but courageous character whose integral role in this tale is often brushed aside. Despite his pivotal role in Raven's success, Fungus Man remains largely absent from the Haida literary and artistic canon. Initially, it seems as though Raven created Fungus Man purely for his own motives, or that the character was a necessary device to advance the plot. However, Haida reverence for conks reveals that Fungus Man's role was neither haphazard nor arbitrary. After all, Raven subjected all living beings to a rigorous testing process before settling on the unlikely polypore, who turns out to be a hero. While it is tempting to ascribe Fungus Man's triumph to his status as a proverbial underdog who succeeds precisely because "he was the least likely candidate," Haida mycology suggests that Fungus Man was no generic conk.[81] Instead, he was a very special species.

78 Augaitis & Wright, *Charles Edenshaw.*

79 Swanton, *Haida Texts.*

80 Ibid.

81 Browne, *Entering Time*, 99.

Until now, most have been satisfied by the notion that Fungus Man was a non-specific conk. Despite efforts by previous scholars to pinpoint the exact type of polypore that Fungus Man symbolizes, these speculations fall short. Northwest Coast Native art expert Bill Holm believed that Fungus Man was a type of red belted conk (*Fomitopsis mounceae*) commonly found throughout western North America.[82] Writer Colin Browne instead offers artist's conk (*Ganoderma applanatum*), a more compelling polypore because its white underbelly stains brown when scratched.[83] Browne argues that this permanent scarring inspired Raven to etch a face into a conk, thus bringing Fungus Man to life.

Mycologist Paul Stamets proposes Agarikon as a new possibility.[84] Although the species is endemic to Haida territory, it remains neglected in most Fungus Man discourse. This unmistakable, endangered conk grows off of tree branches and trunks, perched high above the forest floor where it hovers like an artificial cloud. In old growth forests, the conk reaches impressive dimension—65 feet off the ground and sometimes more than two feet long by two feet in diameter, weighing nearly 20 pounds. If undisturbed, Agarikon grows an additional ring each year, creating a scalloped mass with a smooth bottom layer.[85] Combined with its otherworldly facade, Agarikon's unique phallic shape and labial layers add to its significance as an intersexed talisman, making the specimen a convincing match for Fungus Man (fig. 1.10).

Amongst the Haida and Tlingit, tree fungi such as Agarikon protected against negativity and were capable of deflecting "evil or malicious thoughts."[86] Small pieces of the conk were hung inside a doorway to "repel" ill will from adversaries or integrated by healers during ceremonies.[87] However, bad luck was known to strike those who dislodged an entire conk from a tree, highlighting the importance of sustainable harvesting and preserving old growth forests. Instead, small chunks were carved off of the conk, which remained growing on the tree. In Native Alaskan communities, there exists a story of a conk, "half a mile wide, that could never be depleted," who demanded gifts whenever

82 Wright, "Hlgas7agaa."

83 Browne, *Entering Time*.

84 Stamets, "Novel Antimicrobials."

85 Hodge, "Agarikon."

86 Blanchette et al, "19th Century Shaman Grave Guardians," 123; Turner & Bell, "The Ethnobotany of the Coast Salish"; Turner, "The Ethnobotany of the Bella Coola"; Turner et al, *Ethnobotany of the Nitinaht*; Compton, "Upper North Wakashan & Southern Tsimshian"; Turner, *Ancient Pathways*; Turner & Cuerrier, "'Frog's Umbrella.'"

87 Compton, "Upper North Wakashan & Southern Tsimshian," 137; Stamets, "Novel Antimicrobials"; Turner & Cuerrier, "'Frog's Umbrella.'"

FIGURE 1.10
Agarikon

portions of its body were removed.[88] Neglecting to leave an offering meant that one would be trapped in the forest forever, indebted to the polypore.[89]

A generous yet vengeful fungus, Agarikon is likewise revered for its protective properties, and was frequently dried and burned to create a cosmetic to defend against mosquitos, sun, and general harm.[90] Perhaps the face paint that Raven dons in Edenshaw's Seattle platter is a nod to the trusty companion sitting beside him. Suddenly, Agarikon's reputation for personal safety and good luck make the fungus an obvious choice for Raven's quest.

The bitter tasting Agarikon also possesses medicinal properties that can treat side effects from smallpox and tuberculosis.[91] The Haida were likely aware of its potential during smallpox epidemics, but the velocity of the

88 Turner & Cuerrier, "'Frog's Umbrella,'" 198.

89 Ibid.

90 Duffek & McLennan, *The Transforming Image*; Turner & Cuerrier, "'Frog's Umbrella.'"

91 Blanchette et al, "19th Century Shaman Grave Guardians"; Stamets, "Antipox Properties"; Hwang et al, "Biological & Chemical Evaluation."

disease left little time to develop widespread Agarikon remedies.[92] During the turn of the century, Indigenous and non-Indigenous medical practitioners utilized Agarikon specimens to alleviate tuberculosis.[93] The conk proved a versatile panacea, used by Northwest First Nations to treat headaches when smoked, ease frostbite and wounds when rubbed on affected areas, consumed as a liquid to cure stomachaches, internal bleeding, or gonorrhea, applied to sore joints, and chewed to alleviate sore throats.[94]

Agarikon's mycelium, the root-like structure from which the conk's body emerges, likewise possesses restorative qualities. The white mycelium mat exists below the tree's bark, where it grows and wraps itself around the trunk from the inside. Eventually, the fruiting body of Agarikon peeps out from a crack in the bark, revealing a layered, bulbous white mass that we know as Fungus Man. The mycelium mat can reach several feet in length and ¼ inch in thickness, forming a pliable, felt-like material. When a tree's bark is stripped, the fungal blanket is easily removed from its interior, resembling a piece of torn parchment paper. Indigenous individuals used sections of the mycelium mat to dress open wounds, where its "styptic qualities" halted bleeding, much like a medicated bandage.[95] An absorbent material with blood-clotting abilities, Agarikon's mycelium was favored by loggers and carvers who "collected these felts and stored them for emergency use" while in the woods.[96] Other uses of Agarikon mycelium include diapers for infants, leading one to wonder if the soft, fibrous fungal quilt doubled as a menstrual hygiene product, shedding new light on Fungus Man's relationship with the chitons.[97]

9 Carving Agarikon for Tlingit Mortuary Use

North of Haida country, the Tlingit exist as a distinct Indigenous nation whose southeastern Alaskan territory sometimes overlaps with that of the Haida. Although the two nations possess unique languages and histories, their geographic proximity has led to significant cultural similarity. Like the Haida, the Tlingit hold Agarikon in high regard, noting its spiritual, protective, and artistic traits. Carved Agarikon effigies were placed near or within Tlingit graves where

92 Stamets, "Antipox Properties," "Antiviral Activity."

93 Compton, "Upper North Wakashan & Southern Tsimshian," 138; Rogers, "Medicinal Mycology."

94 Turner & Cuerrier, "'Frog's Umbrella.'"

95 Hubert, *An Outline*; Blanchette et al, "Extraordinary Fungal Masks," 264.

96 McIntyre, "Indian Paint," 5; Gilbertson, "Wood-Rotting Fungi."

97 Blanchette et al, "Extraordinary Fungal Masks."

FIGURE 1.11
Tlingit healer Aanxudaas Dl'oogudzees (Berner's Bay Jim), holding tools used for healing (1900)
WINTER & POND, UNITED STATES PUBLIC DOMAIN, VIA WIKIMEDIA COMMONS

healers were interred.[98] Thought to ward off "bad spirits or ghosts," Agarikon was sometimes also displayed along the walls or rafters of longhouses where healers tended to patients in spiritual distress.[99]

Endeavoring to identify Agarikon effigies, Robert Blanchette's 1992 and 1997 studies investigated 15 carved Tlingit Agarikon figures in museum collections. Museums previously catalogued the hard, brown objects as wooden carvings. However, Blanchette correctly identified the items as Agarikon through careful sampling and testing, noting that the conk's annual growth rings resembled those of tree rings.[100] Looted by ethnologists in the late 1800s, who noted that "the grave guardians" had been "oiled … to increase their dark color," the carvings were sometimes "ornamented with copper" or paint.[101]

Carved Agarikon effigies were common items in a Tlingit healer's tool kit, where the polypore was whittled to resemble "animal or anthropomorphic figures" imbued with powerful attributes (fig. 1.11).[102] Agarikon's thick, woody body and colossal size made it an optimal carving material. Although the effigy itself is a potent object, its intensity doubled when carved from Agarikon.[103] Measuring just under a foot tall, these figures served as "spirit catchers" to

98 Blanchette et al, "19th Century Shaman Grave Guardians"; Stamets, "Novel Antimicrobials"; Turner, *Ancient Pathways*.
99 Compton "Upper North Wakashan & Southern Tsimshian," 137; Hodge, "Agarikon."
100 Blanchette et al, "19th Century Shaman Grave Guardians."
101 Rogers, "Medicinal Mycology," 31.
102 Blanchette et al, "19th Century Shaman Grave Guardians," 123.
103 Ibid, 122.

absorb dangerous entities or repel them, much like Fungus Man's role in obtaining the chitons.[104] Depicting smiling or "singing" beavers, bears, and humans, some of the effigies exhibit large eyes, "o"-shaped mouths, and round bellies, hinting at their spirit-catching abilities.[105] When viewed in this light, Edenshaw's exaggerated Fungus Man depictions make for an ideal Agarikon grave figure. Other Agarikon effigies are more understated, such as the smiling, crouched bear and a human-esque head with a toothy grin (figs. 1.12–1.13).

The Agarikon effigies and Edenshaw platters display striking parallels, providing evidence that Agarikon held similar, albeit distinct, meanings for Tlingit and Haida communities. After all, visiting a dangerous spiritual realm in hopes of curing a patient was central to the work of a Tlingit shaman, bearing resemblance to Raven and Fungus Man's perilous search for female genitalia in uncertain waters. As a talisman that could cure seemingly any ailment and defend against spiritual defects, "it was only natural that the mystical fungus should accompany" a healer into the afterlife.[106] Placed at the head of a burial to protect the dead from unwanted intruders, Agarikon grave figures cautioned others not to approach or vandalize the space—with anthropologists apparently disregarding these warnings.[107]

Of the 15 known Agarikon figures, most are roughly carved and contain little detail. They diverge from Haida and Tlingit art, known for being ornate and precise, as exemplified by Edenshaw's argillite masterpieces. The Agarikon effigies' "unrefined" or "undisciplined" nature suggests that they belong to "the antithetical realm of the supernatural."[108] Their crude portrayals echo Edenshaw's simple depictions of Fungus Man, lacking detailed anatomy or definition beyond an open mouth, large eyes, and squared teeth.

10 The Future of Agarikon

In conclusion, Agarikon challenges prevailing notions of easily separable "natural versus cultural" histories, as well as other structuralist dichotomies such as male/female, life/death, and sacred/profane.[109] The noble fungus reveals that a gray area exists and that things are not always what they seem, a prospect

104 Ibid.

105 Nicholson, "Some Spiritual Uses"; Rogers, "Medicinal Mycology."

106 Hodge, "Agarikon."

107 Jonaitis, *Art of the Northern Tlingit*; Blanchette et al, "19th Century Shaman Grave Guardians"; Nicholson, "Some Spiritual Uses."

108 Ibid, 123; "Style & Meaning."

109 Saussure, *Course in General Linguistics*; Levi-Strauss, *Structural Anthropology*; Diamond, *In Search of the Primitive*.

FIGURE 1.12
Tlingit grave effigy carved from Agarikon, in the likeness of a bear
Catalog #19/309
PHOTOGRAPH COURTESY OF THE DIVISION OF ANTHROPOLOGY, AMERICAN MUSEUM OF NATURAL HISTORY

FIGURE 1.13
Tlingit grave effigy carved from Agarikon, revealing a human head
Catalog #E/701
PHOTOGRAPH COURTESY OF THE DIVISION OF ANTHROPOLOGY, AMERICAN MUSEUM OF NATURAL HISTORY

that makes many uncomfortable. In addition, Agarikon questions the idea that human culture solely influences human agents, and instead forces us to reconsider the impact of human actions on local ecology.[110] A species instrumental to Northwest Coast Indigenous cultures, yet threatened by environmental practices introduced by settler colonialism, Agarikon occupies a precarious position. As old growth forests become increasingly rare throughout Canada and the United States as a result of logging and climate change, Agarikon is at serious risk of going extinct.[111] Taking a somewhat pessimistic tone, Margaret Ronda refers to the Anthropocene as the "end of nature."[112] As dire as it sounds, Agarikon's unique role in ancient creation stories and mortuary practices begs the (rather unsettling) question: *How do we commemorate nature, even when it ceases to exist?* Indeed, the conk is already severely endangered, leaving us to wonder if its status as a mythical being will increase as Agarikon grows more scarce, or, more cynically, if the fungus will inevitably lose its influence.

However, we must not deny the agency of Agarikon itself. As this chapter suggests, humans are intimately entangled with plant-like creatures, each dependent upon the other for physical and cultural survival. By examining Haida and Tlingit depictions of Agarikon, such as Charles Edenshaw's "Fungus Man" argillite platters and carved mortuary items, the fungus reveals that it is more than just an ordinary conk, highlighting why this particular polypore is so integral to Northwest Coast Native societies. Agarikon is "of particular interest because it was crafted and imparted spiritualistic power through shape and substance, rather than through hallucinogenic or mind-altering means," unlike more typical associations of mushrooms.[113] As this chapter demonstrates, the conk continues to be shrouded in mystery and double entendres, although closer inspection of its Haida and Tlingit cultural contexts shows that Agarikon possesses very real, specific meanings for Indigenous individuals. More than just a modest mushroom, Agarikon is a life giver, healing agent, and signifier of death, responsible for a myriad of cultural interventions along the Northwest Coast.

110 Chakrabarty, "The Climate of History."
111 Blanchette et al, "Fungal Mycelial Mats"; Hodge, "Agarikon."
112 Ronda, "Mourning & Melancholia."
113 Nicholson, "Some Spiritual Uses," 26.

Bibliography

Augaitis, Daina & Robin K. Wright. *Charles Edenshaw*. London, England: Black Dog Publishing, 2013.

Augaitis, Daina, Lucille Bell, Nika Collison, Vince Collison, Robert Davidson, Jacqueline Gijssen, Guujaaw, Marianne Jones, Peter MacNair, Bill Reid, Isabel Rorick, Michael Nicoll Yahgulanaas, & Don Yeomans. *Raven Travelling: Two Centuries of Haida Art*. Seattle, Washington: University of Washington Press, 2006.

Baldwin, Andrew & Bruce Erickson. "Introduction: Whiteness, Coloniality, & the Anthropocene." *Environment & Planning, Society & Space* 38, no. 1 (2020): 3–11.

Bauer, Andrew M., Matthew Edgeworth, Lucy E. Edwards, Erle C. Ellis, Philip Gibbard, & Dorothy J. Merritts. "Anthropocene: Event or Epoch?" *Nature* 597 (2021): 332.

Blackman, Margaret B. *During My Time: Florence Edenshaw Davidson, a Haida Woman*. Vancouver, British Columbia, Canada: Douglas & McIntyre, 1982.

Blanchette, Robert A. "Haploporus Odorus: A Sacred Fungus in Traditional Native American Culture of the Northern Plains." Mycologia 89, no. 2 (1997): 233–240.

Blanchette, Robert A. "Extraordinary Fungal Masks Used by the Indigenous People of North America & Asia." *Fungi* 10, no. 3 (2017): 9–12.

Blanchette, Robert A., Brian D. Compton, Nancy J. Turner, & Robert L. Gilbertson. "19th Century Shaman Grave Guardians Are Carved Fomitopsis Officinalis Sporophores." *Mycologia* 84, no. 1 (1992): 119–24.

Blanchette, Robert A., Deborah Tear Haynes, Benjamin W. Held, Jonas Niemann, & Nathan Wales. "Fungal Mycelial Mats Used as Textile by Indigenous People of North America." *Mycologia* 113, no. 2 (2021): 261–267.

Boas, Franz. *Primitive Art*. New York, New York: Dover Publications, 1927.

Bringhurst, Robert. *A Story as Sharp as a Knife: The Classical Haida Mythtellers & Their World*. Vancouver, British Columbia, Canada: Douglas & McIntyre, 1999.

Bringhurst, Robert & Bill Reid. The Raven Steals the Light. Vancouver, British Columbia, Canada: Douglas & McIntyre, 1996.

Browne, Colin. *Entering Time: The Fungus Man Platters of Charles Edenshaw*. Vancouver, British Columbia, Canada: Talonbooks, 2016.

Chakrabarty, Dipesh. "The Climate of History: Four Theses." *Critical Inquiry* 35, no. 2 (2009): 197–222.

Chittenden, Newton H. *Official Report of the Exploration of the Queen Charlotte Islands, For the Government of British Columbia*. Victoria, British Columbia, Canada: Government of British Columbia, 1884.

Cole, Douglas. *Captured Heritage: The Scramble for Northwest Coast Artifacts*. Vancouver, British Columbia, Canada: Douglas & McIntyre, 1985.

Collison, Nika. "Creating for Culture: Edenshaw's Haida Roots & Cultural Transformations." In *Charles Edenshaw*, edited by Daina Augaitis & Robin K. Wright, 21–27. London, England: Black Dog Publishing, 2013.

Compton, Brian D. "'It Pulls Everything to You': North Wakashan Herbal Talismans." Paper presented at the 26th International Conference on Salish & Neighboring Languages, University of British Columbia, Vancouver, British Columbia, Canada, 1991.

Compton, Brian D. "Upper North Wakashan & Southern Tsimshian Ethnobotany: The Knowledge & Usage of Plants & Fungi Amongst the Oweekeno, Hanaksiala (Kitlope & Kemano), Haisla, (Kitamaat) & Kitasoo Peoples of the Central & North Coasts of British Columbia." PhD. dissertation, University of British Columbia, 1993.

Compton, Brian D., Gaston Guzman, & Rolf Mathewes. "Puffballs from the Past: Identification of Gasteromycetes from a Lillooet Archaeological Site & Speculation Regarding Their Aboriginal Use." *Canadian Journal of Archaeology* 19 (1995): 154–159.

Crutzen, Paul J. & Eugene F. Stoermer. "The 'Anthropocene.'" *Global Change Newsletter* 41 (2000): 17–18.

Davidson, Robert. "Reclaiming Haida Culture." In *The Spirit Within: Northwest Coast Native Art from the John H. Hauberg Collection*, edited by Helen Abbott, Steven Brown, Lorna Price, & Paula Therman, 93–99. Seattle, Washington: Seattle Art Museum, 1995.

Davis, Heather & Zoe Todd. "On the Importance of a Date, Or, Decolonizing the Anthropocene." *ACME: An International Journal for Critical Geographies* 16, no. 4 (2017): 761–780.

Dawson, George M. *Report on the Queen Charlotte Islands*. Montreal, Quebec, Canada: Dawson Brothers, 1880.

De Loughrey, Elizabeth M. *Allegories of the Anthropocene*. Durham, North Carolina: Duke University Press, 2019.

Diamond, Stanley. *In Search of the Primitive: A Critique of Civilization*. New Brunswick, New Jersey: Transaction Books, 1974.

Doughty, Christopher E., Adam Wolf, & Christopher B. Field. "Biophysical Feedbacks Between the Pleistocene Megafauna Extinction & Climate: The First Human-Induced Global Warming?" *Geophysical Research Letters* 37, no. 15 (2010): 1–5.

Duffek, Karen & Bill McLennan. *The Transforming Image: Painted Arts of Northwest Coast First Nations*. Vancouver, British Columbia, Canada: Douglas & McIntyre, 2007.

Enrico, John. *Skidegate Haida Myths & Histories*. Skidegate, British Columbia, Canada: Queen Charlotte Islands Museum Press, 1995.

Gagne, Karine. *Caring for Glaciers: Land, Animals, & Humanity in the Himalayas*. Seattle, Washington: University of Washington Press, 2019.

Galloway, Brent. *Upper Sto:lo Ethnobotany*. Sardis, British Columbia, Canada: Coqualeetza Education Training Center, 1982.

Gibbard, Philip, Michael Walker, Andrew Bauer, Matthew Edgeworth, Lucy Edwards, Erle Ellis, Stanley Finney, Jacquelyn L. Gill, Mark Maslin, Dorothy Merritts, & William Ruddiman. "The Anthropocene as an Event, Not an Epoch." *Journal of Quaternary Science* 37, no. 3 (2022): 395–399.

Gilbertson, Robert L. "Wood-Rotting Fungi of North America." *Mycologia* 72 (1980): 1–49.

Govindrajan, Radhika. *Animal Intimacies: Interspecies Relatedness in India's Central Himalayas*. Chicago, Illinois: University of Chicago Press, 2018.

Guernsey, Paul. "The Infrastructures of White Settler Perception: A Political Phenomenology of Colonialism, Genocide, Ecocide, & Emergency." *Environment & Planning, Nature & Space* 5, no. 2 (2022): 588–604.

Haraway, Donna. "Anthropocene, Capitalocene, Chthulhocene." In *Art in the Anthropocene: Encounters Amongst Aesthetics, Politics, Environments, & Epistemologies*, edited by Heather Davis & Etienne Turpin, 255–270. London, England: Open Humanities Press, 2014.

Hodge, Kathie T. 2010. "Agarikon." *Cornell Mushroom Blog. Cornell University*. April 12. https://blog.mycology.cornell.edu/2010/04/12/agarikon/.

Hoover, Alan L. "Charles Edenshaw & the Creation of Human Beings." *American Indian Art Magazine* 8, no. 3 (1983): 62–67.

Hubert, Earnest E. *An Outline of Forest Pathology*. New York, New York: John Wiley & Sons, 1931.

Hwang, Changhwa, B.U. Jaki, Larry L. Klein, David C. Lankin, James Mcalpine, Jose G. Napolitano, S.G. Franzblau, Sanghyun Cho, Paul Stamets, & Guido Pauli. "Biological & Chemical Evaluation of Anti-TB Coumarins From the Polypore Mushroom, Fomitopsis Officinalis." *Planta Medica* 78, no. 11 (2012).

Jenness, Diamond. *The Saanitch Indians of Vancouver Island*. Unfinished manuscript. Available at the Department of Anthropology, Royal British Columbia Museum, Victoria, British Columbia, Canada, 1945.

Jonaitis, Aldona. "Style & Meaning in the Shamanic Art of the Northern Northwest Coast." In *The Box of Daylight: Northwest Coast Indian Art*, edited by Bill Holm, 129–131. Seattle, Washington: University of Washington Press, 1983. *Art of the Northern Tlingit*. Seattle, Washington: University of Washington Press, 1986.

Kari, Priscilla Russell. *Tanaina Plantlore. Dena'ina K'et'una: An Ethnobotany of the Dena'ina Indians of Southcentral Alaska*. Anchorage, Alaska: National Park Service, 1987.

Keeler, Kyle. "Before Colonization (BC) & After Decolonization (AD): The Early Anthropocene, the Biblical Fall, & Relational Pasts, Presents, & Futures." *Environment & Planning, Nature & Space* 5, no. 3 (2021): 1341–1360.

Kimmerer, Robin Wall. *Braiding Sweetgrass: Indigenous Wisdom, Scientific Knowledge, & the Teachings of Plants*. Minneapolis, Minnesota: Milkweed Editions, 2013.

Lachler, Jordan. *Dictionary of Alaskan Haida*. Juneau, Alaska: Sealaska Heritage Institute, 2010.

LaDuke, Winona. *Recovering the Sacred: The Power of Naming & Claiming*. Chicago, Illinois: Haymarket Books, 2005.

Levi-Strauss, Claude. *Structural Anthropology*. Paris, France: Plon Books, 1958.

Lewis, Simon L. & Mark A. Maslin. *The Human Planet: How We Created the Anthropocene*. New Haven, Connecticut: Yale University Press, 2018.

Lightfoot, Kent G., Lee M. Panich, Tsim D. Schneider, & Sara L. Gonzalez. "European Colonialism & the Anthropocene: A View From the Pacific Coast of North America." *Anthropocene* 4 (2013): 101–115.

MacDonald, George F. *Haida Art*. Seattle, Washington: University of Washington Press, 1996.

Marsden, Susan. *Na Amwaaltga Ts'msiyeen: The Tsimshian, Trade, & the Northwest Coast Economy*. Prince Rupert, British Columbia, Canada: Museum of Northern British Columbia, 1992.

McIntyre, Robert N. "Indian Paint & Quinine Fungus Collected." *Yosemite Nature Notes* 28 (1949): 4–6.

Moore, Jason W. *Anthropocene or Capitalocene? Nature, History, & the Crisis of Capitalism*. Oakland, California: PM Press, 2016.

Nicholson, Michael S. "Some Spiritual Uses of Mushrooms." *Fungi* 2 (2009): 26–27.

Peck, Alexandra M. "Mariners, Makers, Matriarchs: Changing Relationships Between Coast Salish Women & Water." *Open Rivers: Rethinking Water, Place, & Community* 21, no. 1 (2022): 7–29.

Ransom, Kelly Marie. "Art as Negotiation: The Reciprocal Construction of Meanings in the Argillite Carvings of Charles Edenshaw." M.A. thesis, University of British Columbia, 1993.

Rogers, Robert D. "Medicinal Mycology: Agarikon, Ghosts of the Pacific Northwest." *Fungi* 6, no. 4 (2013): 30–33.

Ronda, Margaret. "Mourning & Melancholia at the End of Nature." In *Remainders: American Poetry at Nature's End*, 91–112. Stanford, California: Stanford University Press, 2018.

Saussure, Ferdinand. *Course in General Linguistics*. London, England: Duckworth Books, 1916.

Stamets, Paul. "Novel Antimicrobials From Mushrooms." *HerbalGram* 54 (2002): 2–6. "Antipox Properties of Fomitopsis Officinalis Bond et Singer (Agarikon) from the Pacific Northwest of North America." *International Journal of Medicinal Mushrooms* 7, no. 3 (2005): 495–506.

Stamets, Paul. "Antiviral Activity from Medicinal Mushrooms." United States Patent Application, US205/0276815A1, filed January 6, 2005, published December 15, 2005.

Steffen, Will, Wendy Broadgate, Lisa Deutsch, Owen Gaffney, & Cornelia Ludwig. "The Trajectory of the Anthropocene: The Great Acceleration." *The Anthropocene Review* 2, no. 1 (2015): 81–98.

Swanton, John R. *Haida Texts & Myths: Skidegate Dialect*. Washington, D.C.: Smithsonian Institution, 1905.

Teit, James A. *The Shuswap*. New York, New York: American Museum of Natural History, 1909.

Teit, James A. "Mythology of the Thompson Indians." In *Publications of the Jesup North Pacific Expedition*, edited by Franz Boas, 203–416. New York, New York: G.E. Stechert & Company, 1912.

Turner, Nancy J. "The Ethnobotany of the Bella Coola Indians of British Columbia." *Syesis* 6 (1973): 193–220.

Turner, Nancy J. *Food Plants of British Columbia Indians: Part II, Interior Peoples*. Victoria, British Columbia, Canada: British Columbia Provincial Museum, 1978.

Turner, Nancy J. *Food Plants of Interior First Peoples: Revised Edition*. Vancouver, British Columbia, Canada: University of British Columbia Press, 1997.

Turner, Nancy J. *Plant Technology of British Columbia First Peoples*. Vancouver, British Columbia, Canada: University of British Columbia Press, 1998.

Turner, Nancy J. *Ancient Pathways, Ancestral Knowledge: Ethnobotany & Ecological Wisdom of Indigenous Peoples of Northwestern North America*. Montreal, Quebec, Canada: McGill-Queen's University Press, 2014.

Turner, Nancy J. & Marcus A.M. Bell. "The Ethnobotany of the Coast Salish Indians of Vancouver Island." *Economic Botany* 25, no. 1 (1971): 63–104.

Turner, Nancy J. & Alain Cuerrier. "'Frog's Umbrella' & 'Ghost's Face Powder': The Cultural Roles of Mushrooms & Other Fungi for Canadian Indigenous Peoples." *Botany* 100 (2021): 183–205.

Turner, Nancy J. & Harriet V. Kuhnlein. "Camas (Camassia spp.) & Riceroot (Fritillaria spp.): Two Liliaceous 'Root' Foods of the Northwest Coast Indians." *Ecology of Food & Nutrition* 13 (1983): 199–219.

Turner, Nancy J. & Harriet V. Kuhnlein. *Traditional Plant Foods of Canadian Indigenous Peoples: Nutrition, Botany, & Use*. Amsterdam, The Netherlands: Gordan & Breach Publishers, 1991.

Turner, Nancy J., Harriet V. Kuhnlein, & Keith N. Egger. "The Cottonwood Mushroom (Tricholoma Populinum Lange): A Food Resource of the Interior Salish Indian Peoples of British Columbia." *Canadian Journal of Botany* 65 (1985): 921–927.

Turner, Nancy J., John Thomas, Barry F. Carlson, & Robert T. Ogilvie. *Ethnobotany of the Nitinaht Indians of Vancouver Island*. Victoria, British Columbia, Canada: British Columbia Provincial Museum, 1983.

Turner, Nancy J., Laurence C. Thompson, M. Terry Thompson, & Annie Z. York. *Ethnobotany: Knowledge and Usage of Plants by the Thompson Indians of British Columbia*. Victoria, British Columbia, Canada: Royal British Columbia Museum, 1990.

Welsch, Robert. "Haida Pronouns: Hydaburg Dialect." Paper presented at the 10th International Conference on Salishan Languages, Ellensburg, Washington, August 14–16, 1975.

Whyte, Kyle Powys. "Is it Colonial Déjà Vu? Indigenous Peoples & Climate Injustice." In *Humanities for the Environment: Integrating Knowledges, Forging New Constellations of Practice*, edited by Joni Adamson & Michael Davis, 88–104. New York, New York: Routledge, 2016.

Wildcat, Daniel. *Red Alert! Saving the Planet with Indigenous Knowledge*. Wheat Ridge, Colorado: Fulcrum Publishing, 2009.

Williams-Davidson, Terri-Lynn. "How Raven Gave Females Their Tsaw." In *Charles Edenshaw*, edited by Daina Augaitis & Robin K. Wright, 58–61. London, England: Black Dog Publishing, 2013.

Wright, Robin K. "Hlgas7agaa: Haida Argillite." In *The Spirit Within: Northwest Coast Native Art from the John H. Hauberg Collection*, edited by Helen Abbott, Steven Brown, Lorna Price, & Paula Therman, 131–153. Seattle, Washington: Seattle Art Museum, 1995.

Wright, Robin K. *Northern Haida Master Carvers*. Seattle, Washington: University of Washington Press, 2001.

Zalasiewicz, Jan, Colin N. Waters, Mark Williams, Anthony D. Barnosky, Alejandro Cearreta, Paul Crutzen, Erle Ellis, Michael A Ellis, Ian J. Fairchild, Jacques Grinevald, Peter K. Haff, Irka Hajdas, Reinhold Leinfelder, John McNeill, Eric O. Odada, Clement Poirier, Daniel Richter, Will Steffen, Colin Summerhayes, James P.M. Syvitski, Davor Vidas, Michael Wagreich, Scott L. Wing, Alexander P. Wolfe, An Zhisheng, & Naomi Oreskes. "When Did the Anthropocene Begin? A Mid-20th Century Boundary Level is Stratigraphically Optimal." *Quaternary International* xxx (2014): 1–8.

Zalasiewicz, Jan, Mark Williams, Alan Smith, Tiffany L. Barry, Angela L. Coe, Paul R. Brown, Patrick Brenchley, David Cantrill, Andrew Gale, Philip Gibbard, F. John Gregory, Mark H. Hounslow, Andrew C. Kerr, Paul Pearson, Robert Knox, John Powell, Colin Waters, John Marshall, Michael Oates, Peter Rawson, & Philip Stone. "Are We Now Living in the Anthropocene?" GSA (*Geologic Society of America*) *Today* 18, no. 2 (2008): 4–8.

CHAPTER 2

How to Coexist: the Teaching of the *Plant Kingdom* at the Universities of Coimbra and Salamanca (XVIII–XIX)

Carlos Fernando Teixeira Alves

In the 1770s, a pivotal moment occurred at the Universities of Coimbra and Salamanca that would shape the institution's academic landscape for decades to come. In the case of Coimbra, the introduction of new Statutes, meticulously divided into three volumes—Theology, Law, and Natural Faculties (Medicine, Philosophy, and Mathematics)[1]—led this university to embark on a transformative journey. While these statutes covered various aspects of student life, teaching methods, and administrative regulations, it is the pedagogical and scientific component that beckons our work.

Book III of the Statutes unveils a remarkable vision put forth by Portuguese reformers (in this work meaning the politicians and professors who think, support, and apply the changes at the universities at study). It declares that the pursuit of scientific knowledge should revolve around the extensive study of the *Plant Kingdom* (defined, literally, as the study of plants), unraveling the myriad uses and benefits that nature has bestowed upon humanity.

The reformers emphasized the importance of comprehending general plant science, encompassing shared observations and truths across all plant species. Subsequently, they sought to delve into the unique characteristics and applications of each specific plant, particularly highlighting their relevance to the arts and society. However, their focus remained steadfastly utilitarian, emphasizing usefulness while briefly acknowledging the curiosity surrounding certain plant species, "that by observation has been discovered in the different species of Plants, that Nature produces abundantly for the use of Man."[2]

1 *Estatutos da Universidade de Coimbra* (1772), vol. Livro I: Do curso theologico (Coimbra: Por Ordem da Universidade, 1972); *Estatutos da Universidade de Coimbra* (1772), vol. Livro II: Cursos juridicos das Faculdades de canones e de leis (Coimbra: Por Ordem da Universidade, 1972); *Estatutos da Universidade de Coimbra* (1772), vol. Livro III: Cursos das sciencias naturaes e filosoficas (Coimbra: Por Ordem da Universidade, 1972).

2 *Estatutos da Universidade de Coimbra* (1772), 1972, Livro III: Cursos das sciencias naturaes e filosoficas:239–44.

 | DOI:10.1163/9789004721777_004

The politician's and professors' dedication to understanding the human-nature relationship became evident in subsequent sections of the 1772 statutes. Recognizing the limitations of preserving only dried and lifeless plant specimens within the Natural History Cabinet, they advocated for the establishment of a Botanical Garden. This living repository of plants would allow students and scholars to witness the vibrant, thriving nature firsthand, thereby enhancing their comprehension of the natural world:[3]

> to worthily fix the Study of Nature in the Center of the University ... there will be a Room with the capacity that requires a Museum, or Office ... And it will be divided into three compartments, each one for the products of one of the kingdoms of nature.

The University of Coimbra, driven by an ambitious and often utopian vision,[4] initiated radical changes throughout the institution following the 1772 reform. Notably, their endeavors to promote practical subjects and innovative teachings,[5] particularly in philosophy, mathematics, and medicine, underscored a deep-rooted commitment to expanding knowledge:

> the establishment of the natural sciences that was made at the University would make possible a better knowledge of the natural riches existing in the country, bringing new material resources to industry, with the consequent development of trade ... The teaching of the natural sciences implemented could not fail to have as a reflection the development of new arts, new manufactures, new factories and the improvement of existing ones.[6]

These disciplines are now regarded as "useful"[7] experienced unprecedented prominence in university curricula, permeating academia and society at

---

3 *Estatutos da Universidade de Coimbra* (1772), Livro III: Cursos das sciencias naturaes e filosoficas:264–65.

4 Carlos Fernando Teixeira Alves, «A Ordem Natural nas reformas universitárias de Salamanca e Coimbra (1769–1820)» (Tese de Doutoramento em História, Lisboa, Universidade de Lisboa, 2021), 25.

5 A.M. Amorim da Costa, «As Ciências Naturais na Reforma Pombalina da Universidade—"Estudo de Rapazes, não ostentaçã de Príncipes"», in *O Marquês de Pombal e a Universidade* (Coimbra: Imprensa da Universidade de Coimbra, 2014), 179–208.

6 Costa, 207–8.

7 José Luís Cardoso, "Natural law, natural history and the foundations of political economy", in *The Elgar Companion to Economics and Philosophy* (Edward Elgar Pub, 2004), 14.

large. Natural history, as an exemplar, assumed a pivotal role both within and beyond the university, aspiring to facilitate physical and social progress among populations.[8]

The field of philosophy witnessed notable transformations, most notably in the teaching of natural philosophy,[9] which received a significant boost through the introduction of new subjects. The second half of the 18th century witnessed the integration of several "utilitarian" disciplines into European universities.[10] While my focus lies primarily on Coimbra and Salamanca, I shall not delve into an exhaustive analysis of each discipline. Instead, I shall concentrate on the most emblematic and representative subjects that exemplify the new vision of nature and its resources, shared by both Iberian reformers and numerous esteemed professors. Thus, my chapter centers on the disciplines of natural history, experimental physics, chemistry, botany, agriculture, and foronomy (mainly). It is crucial to analyze these disciplines collectively, for they were designed to complement one another, to be taught in a specific order, and to engage in dialogue. Together, they constitute an earnest endeavor to understand and harness nature's resources through rationality and scientific inquiry, thereby forging novel pathways for the exploitation of natural resources.

My central question, therefore, revolves around how the formulation of these selected disciplines illuminates a particular perception of nature and seeks to establish a new relationship between humans, nature, and natural resources. My hypothesis posits that these innovative curricular introductions aimed to leverage the abundant natural resources within the Portuguese and Spanish Empires, to foster economic, industrial, and agricultural development,

8 Cardoso, 15; Emma Spary, "The 'Nature' of Enlightenment", in *The Sciences in Enlightened Europe* (Chicago and London: University of Chicago Press, 1999), 179.

9 William A. Wallace, "Traditional natural philosophy", in *The Cambridge History of Renaissance Philosophy* (Cambridge: Cambridge University Press, 2008), 201–35.

10 Robert D. Anderson, *European universities from the Enlightenment to 1914* (Oxford: Oxford University Press, 2004); Laurence Brockliss, "Curricula", in *A History of the University in Europe: Universities in Early Modern Europe (1500–1800)*, vol. 2 (Cambridge: Cambridge University Press, 2003), 565–620; François Cadilhon, Jean Mondot, and Jacques Verger, *Universités et Institutions Universitaires Européenes au XVIIIe Siècle. Entre Modernisation et Tradition* (Bordeaux: Presses Universitaires de Bordeaux, 1999); Willem Frijhoff, «Modelos», in *Uma História da Universidade na Europa. As Universidades na Europa Moderna (1500–1800)*, vol. 2 (Lisboa: Imprensa Nacional-Casa da Moeda, 2002), 39–105; Notker Hammerstein, «O Iluminismo», in *Uma História da Universidade na Europa. As Universidades na Europa Moderna (1500–1800)*, vol. 2 (Lisboa: Imprensa Nacional-Casa da Moeda, 2002), 595–614; Olaf Pedersen, «Tradição e Inovação», in *Uma História da Univeridade na Europa*, vol. II As Universidades na Europa Moderna (1500–1800), 2002, 435–72.

ultimately enhancing the lives of the respective populations. However, consequently, this exhaustive pursuit triggered profound impacts on biodiversity and ecosystems.

To unravel these historical developments, my research draws upon primary sources that shed light on the pedagogical and scientific aspects of the universities of Coimbra and Salamanca. The cornerstone of my investigation lies in the study plans of these esteemed institutions, spanning multiple years, such as those from 1771, 1807, and 1820 for Salamanca and the 1772 statutes for Coimbra (supplemented by subsequent decrees until 1836). Additionally, I will weave together insights gleaned from professors, scientists, and specialists of the time, who expressed their perspectives through extensive writings, opinions, and testimonials.

1 The Syllabus as a "Map" of the Human-Nature Relationship?

In recent years, the significant impact of human activity on the Earth's ecosystems is one of the main subjects in academia and society in general. One of the most important consequences of this is the rise of the Anthropocene studies—which increasingly provide scientific evidence on the magnitude and scope of human influence on the planet.

Paul Crutzen, a Nobel laureate in chemistry, and Eugene Stoermer, an American geologist,[11] definitively coined the term "Anthropocene;" proposing that human activity had become the dominant force shaping the Earth's ecosystems and geology and that this warranted the recognition of a new geological epoch. Their contribution was the first step towards a new way of thinking about the relationship between humans and the natural world. In the early years of the Anthropocene studies, the focus was primarily on documenting and quantifying the human impact on the Earth's ecosystems and geology. Scientists from a wide range of disciplines, including geology, ecology, and climate science, began to collect and analyze data to understand the scope and scale of human impact on the planet.

This research in the early years of the Anthropocene studies included the impact of human activities on biodiversity, land use change, and pollution. Scientists documented the rapid decline of many plant and animal species, the conversion of natural habitats into urban and agricultural landscapes,

11 Eugene F. Stoermer e Paul Crutzen, "The 'Anthropocene'", *IGBP Newsletter*, n.º 41 (2000): 17–18; Paul Crutzen, "Geology of mankind", *Nature* 415, n.º 23 (2002): 23, https://doi.org/10.1038/415023a.

and the widespread contamination of air, water, and soil with pollutants. As William Ruddiman show,[12] human activities such as deforestation and agriculture have had a significant impact on the Earth's climate since the beginning of the Holocene epoch, long before the Industrial Revolution.

As the Anthropocene studies progressed, scientists began to explore the complex interactions between human activities and the Earth's ecosystems. They recognized that human impact on the planet was not simply a matter of individual actions or isolated incidents, but rather a complex web of interconnected systems. This idea is linked to the definition of the concept of planetary boundaries as Johan Rockström understands it.[13] He has argued that there are nine key processes or systems that are critical for maintaining the Earth's stability and resilience and that humans have already crossed four of these boundaries.

Other lines in the Anthropocene studies began to explore the social and cultural dimensions of human impact on the planet. They recognized that human behavior, values, and beliefs were important drivers of environmental change and that understanding these factors was essential for developing effective strategies for mitigating and adapting to the impacts of human activity on the planet. Timothy Morton[14] argued that the Anthropocene represents a fundamental shift in our understanding of the world and that we need to develop new ways of thinking and being to address the challenges posed by environmental change. His contribution has been to highlight the need for new ethics and politics based on interconnectedness and interdependence. This approach was linked to more ethical and moral dimensions of human impact on the planet.[15] Researchers are examining how human behavior,

---

12 William F. Ruddiman, *Plows, Plagues, and Petroleum: How Humans Took Control of Climate* (Princeton: Princeton University Press, 2017).

13 Johan Rockström et al., "Planetary boundaries: Exploring the safe operating space for humanity", *Ecology and Society* 14(2): 32 (2009), http://www.ecologyandsociety.org/vol14/iss2/art32/.

14 Timothy Morton, *The Ecological Thought* (Harvard University Press, 2010); Timothy Morton, *Hyperobjects: Philosophy and Ecology after the End of the World* (Univ Of Minnesota Press, 2013); Timothy Morton, *Dark Ecology: For a Logic of Future Coexistence* (Columbia University Press, 2016).

15 Morton, *The Ecological Thought*; Morton, *Hyperobjects: Philosophy and Ecology after the End of the World*; Morton, *Dark Ecology: For a Logic of Future Coexistence*; Naomi Oreskes and Erik M. Conway, *Merchants of Doubt: How a Handful of Scientists Obscured the Truth on Issues from Tobacco Smoke to Global Warming* (Bloomsbury Publishing, 2011); Brian Thomas Swimme and Mary Evelyn Tucker, *Journey of the Universe* (Yale University Press, 2014); Mary Evelyn Tucker, *The Emerging Alliance of Religion and Ecology* (University of Utah Press, 2014).

values, and beliefs are shaped by environmental factors and how these factors influence our collective response to environmental challenges.

Although, after all these approaches, we can ask how these analyses can contribute to this intense dialogue about the impact of human activity on nature. The university curricula introduced in Coimbra and Salamanca show us an important shift provided by the Iberian governments about the importance of natural resources.[16] However, this new vision of nature[17] sought to create answers for the problems of the economic backwardness of these countries[18]—in this sense, we witnessed the rise of physiocracy.[19] And the education (in this sense, the training provided by the universities) was indeed a way for states to try to increase their prosperity. For example, the case of the study plan for the University of Salamanca of 1820 reflects this idea well. It associates the practice of teaching agriculture with the needs of commercial and industrial development; and, not least "to the present state of the Nation", which through the study of soils and their relationship with crops

16 Alves, «A Ordem Natural nas reformas universitárias de Salamanca e Coimbra (1769–1820)»; Carlos Fernando Teixeira Alves, "Knowledge, economy, and university in the south of Europe at the end of the eighteenth century. The case of Salamanca and Coimbra", in *La economia della conoscenza: innovazione, produttività e crescita economica nei secoli XIII–XVIII/The knowledge economy: innovation, productivity and economic growth, 13th to 18th century* (Florence: Firenze University Press, 2023), 159–75; Carlos Fernando Teixeira Alves, "Knowledge and Society. The role of two universities in southern Europe in the early nineteenth century: the case of Coimbra and Salamanca", in *History of Intellectual Culture*, vol. 2/2023 (Berlin, Boston: De Gruyter Oldenbourg, 2023).

17 Spary, "The 'Nature' of Enlightenment"; Michel Albin, *L'idée de nature en France dans la primière moitié di XVIIIe sciècle* (Paris: S.E.V.P.E.N., 1963); Pedro Calafate, *A ideia de Natureza no Século XVIII em Portugal (1740–1800)* (Lisboa: Imprensa Nacional-Casa da Moeda, 1994); Robin George Collingwood, *The Idea of Nature* (London; Oxford; New York: Oxford University Press, 1960).

18 Peter Burke, "Fables of the bees: a case-study on views of nature and society", in *Nature and Society in Historical Context* (Cambridge: Cambridge University Press, 1997), 112–23.

19 Eduardo Escartín and Francisco Velasco Morente, "Quesnay y los conceptos generales de la Fisiocracia", in *Ilustración, ilustraciones*, vol. 3 (Real Sociedad Bascongada de Amigos del País: Sociedad Estatal de Conmemoraciones Culturales (SECC), 2009), 275–88; Ernest Lluch, «La Difusión del cameralismo y de la fisiocracia a través de Europa y en especial de España durante el siglo XVIII», *Sapere aude: el «atrévete a pensar» en el siglo de las luces*, n.º 3 (1996): 17–26; Alice J. Araujo Lobo, «Cameralismo: un antecedente más en la evolución de la administración como ciencia», *Visión Gerencial*, n.º 1 (2015): 165–80; Alexandre Mendes Cunha, "Polizei and the System of Public Finance: Tracing the Impact of Cameralism in Eighteenth-Century Portugal", in *The Dissemination of Economic Ideas* (Edward Elgar Publishing, 2011), 65–86; Lluch, «La Difusión del cameralismo y de la fisiocracia a través de Europa y en especial de España durante el siglo XVIII»; Ernest Lluch, «El cameralismo en España», *Economía y economistas españoles* 3 (La Ilustración) (1999): 721–28.

would be of great benefit to "fertile Spain".[20] In this case, the essential changes that happened in the courses of philosophy in the eighteenth century were fundamental for the development of the study of several disciplines and, consequently, the impact of human activity in nature.

In my approach, I stake the idea that the creation of philosophy courses in eighteenth-century universities can be seen as a precursor to the (development of environmental philosophy and) the intense exploitation of nature and natural resources through the new knowledge and know-how causing the environmental impacts that today we relate to the Anthropocene Era.

The Enlightenment thinkers who helped to shape the study of philosophy in this period were interested in questions about the natural world, and their work laid the groundwork for the emergence of environmental ethics in the 20th century. Simultaneously, the development of philosophy courses in the eighteenth century also helped establish the importance of interdisciplinary approaches to studying the natural world.[21] Philosophy courses contain several subjects such as natural sciences, physics, chemistry, and biology, which helped to foster a more integrated and holistic approach to understanding the natural world. In the 18th century, the Enlightenment brought about a renewed interest in the study of nature and the natural world, which had a significant impact on the development of philosophy. Philosophers such as Jean-Jacques Rousseau and Immanuel Kant were deeply interested in questions about the relationship between humans and nature,[22] and their work helped to lay the groundwork for the emergence of environmental ethics in the 20th century.

These new courses of philosophy were highly complex and pondered interactions between humans and the environment and require a nuanced understanding of how social, political, and economic factors intersect with ecological systems and processes. The development of interdisciplinary

---

20 *Informe de la Universidad de Salamanca. Sobre Plan de Estudios, ó sobre su fundacion, altura y decadencia, y sobre las mejoras de que es susceptible: con cuyo motivo presenta un proyecto de Ley sobre la Instruccion Publica* (Salamanca: En la Imprenta de Don Vicente Blanco, 1820), 28 and 56.

21 William A. Wallace, "Traditional natural philosophy", in *The Cambridge History of Renaissance Philosophy* (Cambridge: Cambridge University Press, 2008), 201–35.

22 For this question see: Holmes Rolston, *Environmental Ethics: Duties to and Values in The Natural World* (Temple University Press, 1989); Norton, Bryan G., *Sustainability: a philosophy of adaptive ecosystem management* (Chicago: Univ. of Chicago Press, 2007); Bowler, Peter J., *The Norton History of the Environmental Sciences* (WW Norton & Co Inc, 1993); Callicott, J. Baird, *In Defense of the Land Ethic. Essays in Environmental Philosophy* (State University of New York Press; Eighth Printing edition, 1989).

approaches to the study of the natural world, which was fostered in part by the creation of philosophy courses in the eighteenth century, has helped to lay the groundwork for the study of complex challenges between Humans and Nature. The nuanced understanding of the interactions between social, political, economic factors, and ecological systems, fostered by these courses, has contributed to the interdisciplinary approach needed to tackle the environmental issues we face today.

2 The Power of Plants

The intricate relationship between humans and plants has captivated the attention of scholars across various scientific fields, including botany, ecology, and ethnobotany.[23] However, in recent years, there has been a growing interest in the study of human-plant symbiosis, which refers to the mutually beneficial relationship between humans and plants. This relationship takes many forms, such as the use of plants for food, medicine, and other resources, as well as the cultivation of plants for aesthetic and recreational purposes. The study of human-plant symbiosis highlights the interconnectedness of human and natural systems and underscores the need for more holistic and interdisciplinary approaches to environmental and social issues.

The study of plants has become increasingly relevant in the context of global capitalism,[24] which encompasses the economic system that has emerged in the wake of globalization. Plants have historically played a vital role as commodities in global trade, ranging from widely consumed products like coffee and tea to lucrative crops such as cotton and rubber. However, in recent years, there has been a growing interest in the study of the global trade in plants, including the production and distribution of herbal remedies, exotic foods, and luxury goods. By examining plants through this lens, researchers gain a unique vantage point to explore the social, economic, and environmental

23 Robin Wall Kimmerer, *Braiding Sweetgrass. Indigenous Wisdom, Scientific Knowledge and the Teachings of Plants* (Penguin, 2020); Anna Lowenhaupt Tsing, *The Mushroom at the End of the World: On the Possibility of Life in Capitalist Ruins* (Princeton University Press, 2021); Eduardo Kohn, *How Forests Think: Toward an Anthropology Beyond the Human* (Univ. Of California Press, 2013); Arturo Escobar, *Encountering Development: The Making and Unmaking of the Third World* (Princeton University Press, 2011).

24 For this subject saw Richard Drayton, *Nature's Government: Science, Imperial Britain, and the 'Improvement' of the World* (Yale University Press, 2000); Beckert Sven, *Empire of Cotton: A Global History* (Vintage, 2015); Harriet Ritvo, *The Dawn of Green: Manchester, Thirlmere, and Modern Environmentalism* (University of Chicago Press, 2012).

impacts of global capitalism, while also contemplating alternative models of production and trade that prioritize sustainability.

To understand the role and importance of plants to Iberian reformers, it is essential to examine the study plans implemented at the University of Coimbra and the University of Salamanca. Across the disciplines analyzed, the study plans primarily focus on two central ideas: the utility of natural resources and the significance of plants.

In the field of medicine, students were required to develop an in-depth understanding of plants, particularly those with pharmaceutical potential.[25] Casimiro Gómez Ortega, a renowned professor, and physician, emphasized (in his opinion about the subject of materia medica) the importance of studying plants that could be commercialized for food and medicine.[26] This discipline and its associated botanical knowledge played a decisive role, intertwining botany and agriculture into a comprehensive understanding. The 1807 study plan for the University of Salamanca exemplifies this focus on plants in *materia medica*, emphasizing the need "to know the classes of plants, especially those of daily use in medicine, as well as their qualities."[27]

Similarly, at the University of Coimbra, the study of plants held great importance. In addition to disciplines dedicated to understanding flora, the statutes mandated the establishment of a Botanical Garden in 1772, showcasing various plant species and highlighting their virtues and medicinal uses ("to show the virtues and medicinal uses, that in the said products, have been discovered").[28] However, the significance of plants extended beyond medicine and pharmaceuticals to encompass their role in commerce. Consequently, the University of Coimbra made substantial investments in acquiring books on botany, natural history, and medical subjects, including works by renowned scholars such as Carl Linnaeus and the Count of Buffon.[29] These texts provided invaluable

25 Paula Findlen, "Courting nature", in *Cultures of Natural History* (Cambridge: Cambridge University Press, 1996), 58–59; Harold J. Cook, "Physicians and natural history", in *Cultures of Natural History* (Cambridge: Cambridge University Press, 1996), 91.

26 Archivo Historico Nacional de Madrid (AHN), Consejos, Universidades, Legajos 5459-nº6.

27 Santiago José García Hernández, *La Enseñanza de la Medicina en la Universidad de Salamanca (1808–1833)* (Salamanca: Ediciones Universidad de Salamanca y Santiago José García Hernández, 2016), 68; Archivo de la Universidad de Salamanca (AUSA), Actas de Claustros y Juntas de la Universidad, 1798–1800, AUSA 253, 41r.–50v.

28 *Estatutos da Universidade de Coimbra (1772)*, vol. Livro III: Cursos das sciencias naturaes e filosoficas (Coimbra: Por Ordem da Universidade, 1972), 28.

29 José Antunes, «Notas sobre o sentido ideológico da Reforma Pombalina», *Revista de História das Ideias* 4, T. II (1982): 158.

insights into useful plants and optimal approaches for enhancing their commercial exploitation.[30]

While the importance of plants is evident in these study plans, explicit concerns about plant extinction and the limits of natural resources are not addressed in the analyzed sources. However, examining the reformers' approach to plant studies and the question of global capitalism can shed light on their perspectives. The intersection of plant studies and agricultural systems, as seen in the study plans of Coimbra and Salamanca, offers valuable insights into the historical and ecological dimensions of global capitalism.[31]

Two crucial testimonies reinforce this connection. Pedro Rodríguez de Campomanes, an influential proponent of the new statutes for the University of Salamanca in 1771 and the driving force behind this reform, underscored the practical utility of nature knowledge. He emphasized the precision of mechanical operations and the arts that fostered industry, navigation, tactics, and the various artificial products of commerce within the discipline of mathematics ("the exactitude of mechanical operations ... those arts, which promote industry, navigation, tactics, and all the artificial productions of commerce").[32] Similarly, in 1824, Avelar Brotero, a philosophy professor in Coimbra, recognized the considerable progress made in disciplines such as zoology, botany, and mineralogy, which collectively constituted the realm of natural history. Brotero, aligning with his European counterparts, acknowledged the fundamental nature of these disciplines, for "their great usefulness in medicine, agriculture, commerce, and the arts".[33]

Hence, within the context of Iberian reformers and university study plans, the importance of plants is evident in disciplines in the courses of medicine, philosophy, and mathematics. The study of plants for pharmaceutical and commercial purposes was emphasized, leading universities to invest in important resources and knowledge. However, explicit concerns about plant extinction and the limits of natural resources were not addressed in the analyzed sources. Nonetheless, examining plants across disciplines, particularly within agricultural systems, provides valuable historical and ecological insights into the realm of global capitalism. Prominent reformers and professors recognized the practical utility of nature knowledge for various fields, including industry, commerce, medicine, and agriculture.

30 *Estatutos da Universidade de Coimbra* (1772), 1972, Livro III: Cursos das sciencias naturaes e filosoficas:242.

31 Jason W. Moore, *Capitalism in the Web of Life: Ecology and the Accumulation of Capital* (Verso, 2015).

32 Archivo Historico Nacional de Madrid (AHN), Conde de Campomanes, Caixa 34/ Documento 5.

33 Arquivo da Universidade de Coimbra (AUC), Processos de Professores, Cx. 26.

3 The Disciplines That "Unveil" Nature

The curricula of the courses of philosophy, arts, mathematics, and medicine—adopted in Coimbra and Salamanca—gathered the various disciplines that intended to study nature and its resources, thus enhancing a better capacity to exploit them—aiming at the improvement and development of agriculture, industry, and trade, benefiting, as expected, from the resources available in their colonies. In this sense, it was necessary to create a structure of their courses that had two guidelines: from the simplest to the most complex; and, in the case of the disciplines that studied natural products, the students began by studying the various natural resources (flora, fauna, minerals), their composition, purposes, and only then learned the chemical formulas for the transformation of these products—which was performed in laboratories where students were required to have practical classes. Throughout this chapter, I have already mentioned some disciplines introduced in the second half of the 18th century in Portugal and Spain.

However, due to their centrality and importance, we will look more closely at the content of the disciplines of natural history, chemistry, experimental physics, and foronomy. By focusing on the study and comprehension of plants, we can provide important insights into the complex relationships between humans and the environment, particularly in the context of the university reforms of the eighteenth century. These disciplines offer a unique perspective on how human activity has shaped the natural world and, on the challenges, faced in preserving the ecological systems that sustain life on Earth. In the context of the Anthropocene, natural history provides important insights into how human activity has altered ecosystems and disrupted ecological systems. For example, the study of plant populations can reveal how human activities such as deforestation, agricultural expansion, and urbanization have led to the fragmentation and degradation of natural habitats. By studying the distribution and abundance of plant species, researchers can gain a better understanding of the impacts of human activity on biodiversity and ecological health.

It is important to remember that it was from the fragmentation of this chair (natural history), that other chairs such as botany, mineralogy, zoology, and agriculture emerged, which also responded to the need for specialization.[34] In 1772, the chair of the natural history was introduced at the University of

34 Janet Browne, "Natural History", in *The Oxford Companion to the History of Modern Science* (Oxford: Oxford University Press, 2003), 559–63; Phillip R. Sloan, "Natural history, 1670–1802", in *Companion to the History of Modern Science* (London: Routledge, 1996), 295–313.

Coimbra and, as the statutes themselves state, it should be the basis "to physics, and all the arts."[35] In the case of Salamanca, we must wait until the 1807 plan to see the introduction of this discipline[36]—and further development in the 1820 plan.[37] However, the content was similar. In this chair, teachers were to conduct a detailed description of products of vegetable, animal, and mineral origin. The lecturer was then to demonstrate which products had commercial and medicinal purposes.

4 Conclusion

It should be stated that another discipline that had as its objective the study of natural products—mainly linked to the training of physicians—was experimental physics. In the case of Salamanca, the definition of this discipline deserved some attention.[38] However, in 1771 it became present in the new study plan and was intended to make products for medicinal purposes known to the students. This connection with the training of medical students will mark the evolution of this subject and, in 1783,[39] there are already references to a professor of medicine teaching this subject. It is possible to conclude that it would have ceased to be taught from 1807, when the curricular offer increased, and other subjects such as physics and chemistry began to be taught. In Coimbra, the content of this subject was not very different. It already had a strong practical component and was intended to "discover the veil of nature and to ask its most recondite secrets."[40] The teacher had to focus his exposition on the properties of bodies: porosity, compressibility, mobility, and elasticity, among other characteristics.

Agriculture was an essential activity in the Iberian Peninsula, and in the areas that made up the Portuguese and Spanish empires. We also saw that

35 *Estatutos da Universidade de Coimbra* (1772), 1972, Livro III: Cursos das sciencias naturaes e filosoficas:245.

36 George M. Addy, *The Enlightenment in the University of Salamanca* (Califórnia: Duke University Press, 1966), 373.

37 *Informe de la Universidad de Salamanca. Sobre Plan de Estudios, ó sobre su fundacion, altura y decadencia, y sobre las mejoras de que es susceptible: con cuyo motivo presenta un proyecto de Ley sobre la Instruccion Publica* (Salamanca: En la Imprenta de Don Vicente Blanco, 1820), 53.

38 Addy, *The Enlightenment in the University of Salamanca*, 318.

39 Archivo General de Simancas (AGS), Secretaria de Gracia y Justicia. Instrucción Pública. Fechos de la Universidad de Salamanca, 1777–1788, Legajo 945.

40 *Estatutos da Universidade de Coimbra* (1772), 1972, Livro III: Cursos das sciencias naturaes e filosoficas:250.

the attempt to overcome the economic backwardness passed through an improvement in agricultural production and, in this sense, disciplines were introduced in several universities. In the case of Coimbra, it appeared in 1790, and Salamanca in 1820.[41] Mineralogy and botany would have already led students to study soils and plant cultivation. However, with the introduction of this discipline, the study of agricultural practice became systematic and with the support of its infrastructure.

Lastly, the course of mathematics, introduced in Coimbra, and in Salamanca doesn't exist as a course, but just some subjects that were taught in arts and philosophy. In the 1772 statutes for Coimbra, we can saw that math reunited several disciplines, "there follow many other subjects, which have the name of Mixed Mathematics, or *physical mathematics* Sciences … in which many effects and phenomena of Nature are understood". And, as the statutes themselves indicate, one of the most important was the discipline of agronomy (*Foronomia* or Foronomy, the study the laws of equilibrium and motion of bodies): in which is contained the best part of Physics, because movement is the main agent of all phenomena, as the soul of Nature itself. For example, further on, the statutes, indicates some of the various disciplines, where we highlight Hydraulics—important for understanding the hydric currents.[42]

Understanding "fluids" is another recurring theme. But what does this expression mean? It refers to issues that study hydraulic resources. The lecturer was to show the principles of "Hydraulic Architecture; and of the Machines, which have been happily imagined for the conduction, and elevation of waters; for the direction, and distribution of rivers for the benefit of Agriculture, and Fertility of the whole Provinces."[43]

In 1772 chemistry became part of the philosophy course, although it was attended by other students, of medicine and mathematics. It was considered the third part of natural philosophy and was supposed to teach "to separate the different substances which enter into the Composition of a Body; to examine each of its parts."[44] In Coimbra, the students would have to be familiarized

41 Arquivo Nacional da Torre do Tombo (ANTT), Ministério do Reino. Instrução. Requerimentos da Universidade de Coimbra, 1771/1800, Mç. 511, Cx. 637; *Informe de la Universidad de Salamanca. Sobre Plan de Estudios, ó sobre su fundacion, altura y decadencia, y sobre las mejoras de que es susceptible: con cuyo motivo presenta un proyecto de Ley sobre la Instruccion Publica*, 28 and 56.

42 *Estatutos da Universidade de Coimbra* (1772), vol. Livro III: Cursos das sciencias naturaes e filosoficas (Coimbra: Por Ordem da Universidade, 1972), 163.

43 *Estatutos da Universidade de Coimbra* (1772), Livro III: Cursos das sciencias naturaes e filosoficas:186.

44 *Estatutos da Universidade de Coimbra* (1772), Livro III: Cursos das sciencias naturaes e filosoficas:251.

with a vast number of products, of saline, metallic, oily origin, etc.; whereas in Salamanca, the difference was like the products, which were to be directed to medical teaching.

In inference, the study of these disciplines that focus on views on nature and its products can provide important insights into how human activity is shaping the natural world. These disciplines offer unique perspectives on the impacts of human activity on ecological systems and on the challenges in preserving the biodiversity and ecological health of our planet. By studying these relationships and developing more sustainable approaches to managing our natural resources, we can help to ensure a more resilient and sustainable future for ourselves and the natural world.

In conclusion, it should be stated that the university curricula introduced at the University of Coimbra and the University of Salamanca in the second half of the 18th century reflected a new vision of nature and its resources. These reforms aimed to address the economic backwardness of the Iberian countries and promote economic, industrial, and agricultural development. The introduction of practical subjects and the emphasis on disciplines such as natural history, experimental physics, chemistry, agriculture, and foronomy sought to provide the knowledge and know-how necessary for the intensive exploitation of natural resources.

The creation of philosophy courses in the eighteenth century played a crucial role in establishing interdisciplinary approaches to studying the natural world. Philosophers of the Enlightenment, such as Rousseau and Kant, explored questions about the relationship between humans and nature, laying the groundwork for the emergence of environmental ethics in the 20th century. These courses fostered a nuanced understanding of the complex interactions between social, political, and economic factors, and ecological systems, contributing to the interdisciplinary approach needed to address environmental challenges.

Furthermore, the study of human-plant symbiosis and the importance of plants highlighted the interconnectedness of human and natural systems. The relationship between humans and plants has been studied in various scientific fields, including botany, ecology, and ethnobotany. The study of human-plant symbiosis emphasized the mutually beneficial relationship between humans and plants, encompassing their use for food, medicine, resources, and aesthetic purposes. The recognition of the importance of human-plant symbiosis has led to more holistic and interdisciplinary approaches to environmental and social issues.

Additionally, the study plans of the universities focused on the utility of natural resources and the significance of plants. In disciplines such as in medicine and philosophy, students were required to have in-depth knowledge of plants, particularly those important for pharmaceutical purposes and commercial potential. The practical component of the study plans, such as the establishment of the Botanical Garden, emphasized the importance of hands-on experience and the showcasing of the virtues and medicinal uses of plants.

However, it is essential to recognize that while these reforms aimed to enhance economic prosperity, they also had consequences for the environment. The intensive exploitation of natural resources and the emphasis on commercial exploitation led to the depletion of certain species and the impact on ecosystems. The historical roots of the human relationship with nature and the development of new knowledge and know-how must be acknowledged to develop effective strategies for a sustainable future.

Overall, the reforms in the university curricula during the 18th century represented a significant shift in the understanding of nature and its resources. The interdisciplinary approach, the recognition of the interconnectedness of human and natural systems, and the emphasis on the power of plants all played a role in shaping the study plans and the relationship between humans, nature, and natural resources. As we continue to grapple with the environmental challenges of the present, understanding the historical context and the impact of these reforms can guide us in developing sustainable practices and preserving the world we inhabit.

Bibliography

Addy, George M. *The Enlightenment in the University of Salamanca*. Califórnia: Duke University Press, 1966.

Albin, Michel. *L'idée de nature en France dans la primière moitié di XVIIIe sciècle*. Paris: S.E.V.P.E.N., 1963.

Alves, Carlos Fernando Teixeira. «A Ordem Natural nas reformas universitárias de Salamanca e Coimbra (1769–1820)». Tese de Doutoramento em História, Universidade de Lisboa, 2021.

Alves, Carlos Fernando Teixeira. "Knowledge and Society. The role of two universities in southern Europe in the early nineteenth century: the case of Coimbra and Salamanca". In *History of Intellectual Culture*, Vol. 2/2023. Berlin, Boston: De Gruyter Oldenbourg, 2023.

Alves, Carlos Fernando Teixeira. "Knowledge, economy, and university in the south of Europe at the end of the eighteenth century. The case of Salamanca and Coimbra". In *La economia della conoscenza: innovazione, produttività e crescita economica nei secoli XIII–XVIII/The knowledge economy: innovation, productivity and economic growth, 13th to 18th century*, 159–75. Florence: Firenze University Press, 2023.

Anderson, Robert D. *European universities from the Enlightenment to 1914*. Oxford: Oxford University Press, 2004.

Antunes, José. «Notas sobre o sentido ideológico da Reforma Pombalina». *Revista de História das Ideias* 4, T. II (1982): 143–97.

Araujo Lobo, Alice J. «Cameralismo: un antecedente más en la evolución de la administración como ciencia». *Visión Gerencial*, n.º 1 (2015): 165–80.

Archivo General de Simancas (AGS), Secretaria de Gracia y Justicia. Instrucción Pública. Fechos de la Universidad de Salamanca, 1777–1788, Legajo 945.

Archivo Historico Nacional de Madrid (AHN), Consejos, Universidades, Legajos 5459-nº6.

Archivo Historico Nacional de Madrid (AHN), Conde de Campomanes, Caixa 34/ Documento 5.

Arquivo da Universidade de Coimbra (AUC), Processos de Professores, Cx. 26.

Arquivo Nacional da Torre do Tombo (ANTT), Ministério do Reino. Instrução. Requerimentos da Universidade de Coimbra, 1771/1800, Mç. 511, Cx. 637.

Bowler, Peter J. *The Norton History of the Environmental Sciences*. WW Norton & Co Inc, 1993.

Brockliss, Laurence. "Curricula". In *A History of the University in Europe: Universities in Early Modern Europe (1500–1800)*, 2:565–620. Cambridge: Cambridge University Press, 2003.

Browne, Janet. "Natural History". In *The Oxford Companion to the History of Modern Science*, 559–63. Oxford: Oxford University Press, 2003.

Burke, Peter. "Fables of the bees: a case-study on views of nature and society". In *Nature and Society in Historical Context*, 112–23. Cambridge: Cambridge University Press, 1997.

Cadilhon, François, Jean Mondot, and Jacques Verger. *Universités et Institutions Universitaires Européenes au XVIIIe Siècle. Entre Modernisation et Tradition*. Bordeaux: Presses Universitaires de Bordeaux, 1999.

Calafate, Pedro. *A ideia de Natureza no Século XVIII em Portugal (1740–1800)*. Lisboa: Imprensa Nacional-Casa da Moeda, 1994.

Callicott, J. Baird. *In Defense of the Land Ethic. Essays in Environmental Philosophy*. State University of New York Press; Eighth Printing edition, 1989.

Cardoso, José Luís. "Natural law, natural history and the foundations of political economy". In *The Elgar Companion To Economics and Philosophy*, 3–23. Edward Elgar Pub, 2004.

Collingwood, Robin George. *The Idea of Nature*. London; Oxford; New York: Oxford University Press, 1960.

Cook, Harold J. "Physicians and natural history". In *Cultures of Natural History*, 91–105. Cambridge: Cambridge University Press, 1996.

Costa, A.M. Amorim da. «As Ciências Naturais na Reforma Pombalina da Universidade—"Estudo de Rapazes, não ostentaçã de Príncipes"». In *O Marquês de Pombal e a Universidade*, 179–208. Coimbra: Imprensa da Universidade de Coimbra, 2014.

Crutzen, Paul. "Geology of mankind". *Nature* 415, n.º 23 (2002): 23. https://doi.org/10.1038/415023a.

Cunha, Alexandre Mendes. "Polizei and the System of Public Finance: Tracing the Impact of Cameralism in Eighteenth-Century Portugal". In *The Dissemination of Economic Ideas*, 65–86. Edward Elgar Publishing, 2011.

Drayton, Richard. *Nature's Government: Science, Imperial Britain, and the «Improvement» of the World.* Yale University Press, 2000.

Escartín, Eduardo, and Francisco Velasco Morente. «Quesnay y los conceptos generales de la Fisiocracia». In *Ilustración, ilustraciones*, 3:275–88. Real Sociedad Bascongada de Amigos del País: Sociedad Estatal de Conmemoraciones Culturales (SECC), 2009.

Escobar, Arturo. *Encountering Development: The Making and Unmaking of the Third World.* Princeton University Press, 2011.

Estatutos da Universidade de Coimbra (1772). Vol. Livro I: Do curso theologico. Coimbra: Por Ordem da Universidade, 1972.

Estatutos da Universidade de Coimbra (1772). Vol. Livro II: Cursos juridicos das Faculdades de canones e de leis. Coimbra: Por Ordem da Universidade, 1972.

Estatutos da Universidade de Coimbra (1772). Vol. Livro III: Cursos das sciencias naturaes e filosoficas. Coimbra: Por Ordem da Universidade, 1972.

Findlen, Paula. "Courting nature". In *Cultures of Natural History*, 57–74. Cambridge: Cambridge University Press, 1996.

Frijhoff, Willem. «Modelos». In *Uma História da Universidade na Europa. As Universidades na Europa Moderna (1500–1800)*, 2:39–105. Lisboa: Imprensa Nacional-Casa da Moeda, 2002.

García Hernández, Santiago José. *La Enseñanza de la Medicina en la Universidad de Salamanca (1808–1833)*. Salamanca: Ediciones Universidad de Salamanca y Santiago José García Hernández, 2016.

Hammerstein, Notker. «O Iluminismo». In *Uma História da Universidade na Europa. As Universidades na Europa Moderna (1500–1800)*, 2:595–614. Lisboa: Imprensa Nacional-Casa da Moeda, 2002.

Informe de la Universidad de Salamanca. Sobre Plan de Estudios, ó sobre su fundacion, altura y decadencia, y sobre las mejoras de que es susceptible: con cuyo motivo presenta un proyecto de Ley sobre la Instruccion Publica. Salamanca: En la Imprenta de Don Vicente Blanco, 1820.

Kimmerer, Robin Wall. *Braiding Sweetgrass. Indigenous Wisdom, Scientific Knowledge and the Teachings of Plants*. Penguin, 2020.

Kohn, Eduardo. *How Forests Think: Toward an Anthropology Beyond the Human*. Univ. of California Press, 2013.

Lluch, Ernest. «El cameralismo en España». *Economía y economistas españoles* 3 (La Ilustración) (1999): 721–28.

Lluch, Ernest. «a Difusión del cameralismo y de la fisiocracia a través de Europa y en especial de España durante el siglo XVIII». *Sapere aude: el «atrévete a pensar» en el siglo de las luces*, n.º 3 (1996): 17–26.

Moore, Jason W. *Capitalism in the Web of Life: Ecology and the Accumulation of Capital*. Verso, 2015.

Morton, Timothy. *Dark Ecology: For a Logic of Future Coexistence*. Columbia University Press, 2016.

Morton, Timothy. *Hyperobjects: Philosophy and Ecology after the End of the World*. Univ Of Minnesota Press, 2013.

Morton, Timothy. *The Ecological Thought*. Harvard University Press, 2010.

Norton, Bryan G. *Sustainability: a philosophy of adaptive ecosystem management*. Chicago: Univ. of Chicago Press, 2007.

Oreskes, Naomi, and Erik M. Conway. *Merchants of Doubt: How a Handful of Scientists Obscured the Truth on Issues from Tobacco Smoke to Global Warming*. Bloomsbury Publishing, 2011.

Pedersen, Olaf. «Tradição e Inovação». In *Uma História da Univeridade na Europa*, II As Universidades na Europa Moderna (1500–1800):435–72, 2002.

Ritvo, Harriet. *The Dawn of Green: Manchester, Thirlmere, and Modern Environmentalism*. University of Chicago Press, 2012.

Rockström, Johan, Will Steffen, Kevin Noone, Åsa Persson, F. Stuart III Chapin, Eric Lambin, Timothy M. Lenton, et al. "Planetary boundaries: Exploring the safe operating space for humanity". *Ecology and Society* 14(2): 32 (2009). http://www.ecologyandsociety.org/vol14/iss2/art32/.

Rolston, Holmes. *Environmental Ethics: Duties to and Values in The Natural World*. Temple University Press, 1989.

Ruddiman, William F. *Plows, Plagues, and Petroleum: How Humans Took Control of Climate*. Princeton: Princeton University Press, 2017.

Sloan, Phillip R. "Natural history, 1670–1802". In *Companion to the History of Modern Science*, 295–313. London: Routledge, 1996.

Spary, Emma. "The 'Nature' of Enlightenment". In *The Sciences in Enlightened Europe*, 272–304. Chicago and London: University of Chicago Press, 1999.

Stoermer, Eugene F., and Paul Crutzen. "The 'Anthropocene'. *IGBP Newsletter*, n.º 41 (2000): 17–18.

Sven, Beckert. *Empire of Cotton: A Global History*. Vintage, 2015.

Swimme, Brian Thomas, and Mary Evelyn Tucker. *Journey of the Universe*. Yale University Press, 2014.

Tsing, Anna Lowenhaupt. *The Mushroom at the End of the World: On the Possibility of Life in Capitalist Ruins*. Princeton University Press, 2021.

Tucker, Mary Evelyn. *The Emerging Alliance of Religion and Ecology*. University of Utah Press, 2014.

Wallace, William A. "Traditional natural philosophy". In *The Cambridge History of Renaissance Philosophy*, 201–35. Cambridge: Cambridge University Press, 2008.

CHAPTER 3

Tēichigi Colors of Artistic Resistance

James Jack

FIGURE 3.1 Tēichigi from Uncle Yoichiro's garden, shared along with artworks by author at Tokunoshima Library Open Space, March 2024.
PHOTO: OKUBO YUKARI

•••

I love this island

This point in the South Pacific   this point in the Northern hemisphere
And   this point   in the history of the world
I love this point

This island   challenges the Kuroshio Current with resolve
It is our island   for 200,000 of us
I am searching for the lighthouse humans needs to survive here.[1]
Izumi Hōrō

∴

1 Excerpt from poem "Island," *Izumi Hōrō Shinshu*, 23–24. All translations by author.

 | DOI:10.1163/9789004721777_005

1 Vegetal Ancestors

Now banana trees grow abundantly at the site where my kin maternal grandmother's home stood on Tokunoshima (徳之島, "Island of Benevolence"). Cycad palms, rhododendron and bird's nest ferns encircle uncle's house next door in this village of Todoroki. Sugarcane fields expand around all the roads that lead here. Memories of these lush forests live in the excitement of my mother-in-law whenever topics related to plants from her homeland come up in conversation. She glows brightly whenever talking about plants, especially orchids, papayas, hibiscus and others from subtropical climates like the Ryūkyū archipelago. Often these conversations are in contrast to those species that cannot grow in the cold Kantō region around Tokyo where she has spent most of her life since leaving the island at age 16 for work. I have watched her replant papaya trees every year with hope they will fruit over the past decade, despite them shriveling up from frost each winter. Despite adjustments to daily life such as reserved ways of speaking, frigid ways of relating to others and very salty food, memories of the island always remain in vegetal connections. Through her love of plants, a lineage of vegetal ancestors from Ryūkyū will always be part of our family now and for generations to come.

As an artist touching plants softly, I notice their scent, color and behaviors as ways to get to know my ancestors. Starting with this bodily connection to vegetal family members, I am enthralled when their colors expand in contact with

FIGURE 3.2 Toshi Genichiro album #2, "Kyūra ga mēra! Wakya shima ga ichiban dō!" (Warmest greetings! Our island is number one) 2016

FIGURE 3.3 Artwork in progress James Jack studio, 2020

air, water and paper fibers. Plants teach lessons each day to all those who listen. Learning their uses in local cultures for medicine, dyeing and ceremonies along with their names in local dialect helps to understand the self, link with healing cycles and engage in "the muddled journey of decolonization."[2] I work with plants available in the immediate environment according to what is currently in season. Expressing vegetal voices in artistic ways is a decolonial act that recenters island traditions within a resurgence of indigenous cultures[3] occurring today. This is activated through planting, growing, tending, harvesting and composting together with family and friends intertwined. My own story is one of personal family links between islands commencing with Scottish Highland mariner roots and my best friend in kindergarten from Sendai, Japan. From one I learned "the renaissance of indigenous culture implies a serious threat to continued colonial domination."[4] And from another I learned the importance of reciprocal economies of giving. Both are crucial lessons integrated into my creative work that weaves politics with everyday life.

2 Aiko Yamashiro, "Gathered by Plants: Some Decolonial Love Letters," in *Indigenous Pacific Islander Eco-Literatures*, 213.

3 Alfred, 76–81.

4 Hechter, 73.

FIGURE 3.4 James Jack. *Tēichigi Resistance* (*Six/Sun*), 2024. Handmade Yeddo Hawthorn ink on paper. 105 × 105 cm
PHOTO BY JING KAI

2 *Tēichigi* Blood

While peeling fresh bark from tree branches, uncle Yoichiro's hands become stained deep red in the bright July sun. As we strip bark from *Tēichigi*, (テェイチギ, *sharinbai*/車輪梅, Japanese Yeddo Hawthorn, *phaphiolepsis indica var umbellate*)[5] branches freshly cut from his yard, sepia brown bark releases tannins[6] that shift into deep shades of rust red in contact with air. While we peel the bark shards together off the branches, we talk about layers of life: how his mother passing away recently shifts responsibilities, welcoming his son to

5 Kobashigawa, 070.

6 Nakata refers to the plant as "Rhaphiolepis umbellate" and "Japanese Hawthorn" in English. *Shodoshima ko kigitachi*, 100.

TABLE 3.1 Translation of Figure 3.4 into English

Resist	Definitely NO!!	Resist	Resist	Ancestors are Angry!
White Clouds	Single warp yarn, tortoise shell	Rice	Resist Base	Will to Live Gently, Calm, Peacefully
Beautiful Island for Children and Grandchildren	Sun	Intolerant of Government Maneuvering	No Need for Base	All Agricultural Labor Unions
No Need Base Nago City Residents	Dragonfly	NO BASE	Sea Urchin	No USA Relocate Tokunoshima
Total Resistance	Citizens Oppose!!!!	Toy windmill (spinning)	(No) U.S. base relocation in prefecture	Resist Base Resist Base

return to the island to take care of the cattle full-time, and caring for his wife whose fragile body is facing a terminal illness. I share stories of sleepless life during the first year of becoming a parent, the need for decolonial thinking on the island where we live and working through my first university teaching job. In the air we breathe, the ink tones shift deeper towards blood red, while our hands connect with ancestors who have worked with botanical dyes for centuries on Tokunoshima. Feeling their spirits in our bodies while engaging in craft labor, the aphorism "children are ancestors reborn," comes alive in our own family lineage.[7]

Soaked in sun, air and love *Tēichigi* bark shrivels into twisted morsels of auburn red. After a few days in the yard next to uncle's house, I soak the thin strips in a large metal vat of water. Cousin Shuichi has come to visit during one

7 Matsuyama, 143.

FIGURE 3.5 *Sharinbai* evergreen shrub
PHOTO: NIJI NO KAI

of the last university breaks before starting his first job in Tokyo, and eagerly helps to light a fire together while sugarcane *shōchu* cups remain full. As the coals heat up and the water stained with *Tēichigi* bark comes to a boil, we share stories about living in one's own culture and experiences abroad, strain with siblings, as well as the search for mindfulness amidst suffering. While tending to the fire, chestnut pink tones seep from the bark fibers into the vat, only visible with a flashlight in the dark summer starlit night. The extract of *Tēichigi* tree bark has been used as a fabric dye for generations, often infused with mud pigments in the Amami island chain which is part of a *michi no shima* ("pathway of islands") between Taiwan and Kyūshū where cloth, kimono and fabrics have been carried in boats for more than a thousand years.

Plants *are* ancestors; plants *are* siblings. Just as vegetal roots soak up nourishment from the earth, we soak energy from the plants. As one of the youngest species on earth, humans have much to learn from plants and other ancestors.[8] Working caringly with our vegetal friends as "plantcestors"[9] can heal trauma, hurt, and violence. As an artist raised on Turtle Island who has

8 Kimmerer.

9 Discussion with youth workshop participants leads to the invention of a new word "Plantcestors" (Plant + Ancestors) Nelson et al., 0:55:25.

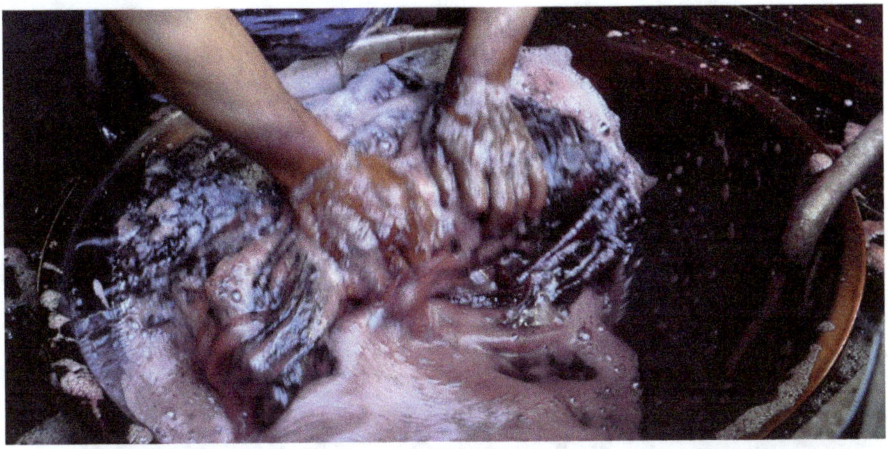

FIGURE 3.6 The dye of *Sharinbai* is enriched with tannin. Silk threads and Kasurimushiro are dyed about 80 times
PHOTO: OSHIMA TSUMUGIMURA

FIGURE 3.7
Sharinbai
DRAWING BY NAKATA MASAGO, 2022

FIGURE 3.8 James Jack. *Tēichigi Resistance* (*Five/Rice*), 2024. Handmade Yeddo Hawthorn ink on paper. 105 × 105 cm
PHOTO BY JING KAI

spent half of his life in the archipelago of Japan, I am entangled with stories of power that enmesh these lands. The US government "waging peace"[10] and Japan's efforts to repeal Article 9 to fortify a military both resuscitate violences of war. I devote my work to raising seeds arising from peaceful friendships and family connections that maintain strength across distances. I hold the proverb, "Treasure *This* Island"[11] in my heart throughout art creation, family relations and livelihood.

10 Part of Eisenhower's term "waging peace" during the onset of the Cold War in the 1950s included the negotiation of a peace treaty that returned Amami and Tokunoshima islands to Japan from US Military control. Eldridge, 31.

11 Matsuyama, 110.

TABLE 3.2 Translation of Figure 3.8 into English

Firm Opposition	Wheel	Tree Stump	Total Resistance	NO BASE
Wisteria	Resist	Resist Base	Anger	Wheel
Oppose Relocation of U.S. Base!	Want to Live Carefree	*Tsuganagajū*	Total Resistance	Lattice
Water Well Pattern	Long Cross	Hey Government, Island Spirit Won't Sell for Money	Resist Base	Coin Pattern
No! Won't Allow Base to be built on Tokunoshima	Sea Urchin	Resist Base	Water well *kasuri* pattern	Firm Dissent

Listen to your vegetal friends nearby. Each rich tone of bark, leaves and nuts has emerged from days of sunlight, wind, rain, and care that protect it within the churning cycles of the forest. The readiness of the plants for interaction is not their "native" origins, or their "genius" owners, but rather indicates the intimate daily connection between human and more than human life tuned into similar frequencies. Recognizing friends beyond the human species together with "indigenous artists against the anthropocene,"[12] observe them in the warm, in the cold and all of the days in between as they shift tones, twist into new shapes, adjust their sheen to shun insects, open and close their vascular networks with the sunlight. Our veins move together with liquids in the

12 Horton, 48–69.

botanic world, attracted to sunlight from within and pulled towards the loving embrace of water. Now seeking a way of life that has been prohibited over centuries is a powerful act of affirming island culture as a contemporary way of being. Roots nourished by ancestors who cared for the plants throughout the seasons provide visible branches, leaves and fruits.

> *"Shimanchu no kokoro wo moteasobu na!"* (*Don't mess around with our islander hearts.*)[13]

3 Archipelagic Solidarity

Observing trees stripped bare alongside American troops standing during an Okinawan land battle during World War II, I feel the presence of densely packed fruit of *Adan* (*Pandanus*) tree fibers alive on the earth's surface. Each nut holds the power for regeneration even in a desolate battlefield. American and Japanese soldiers left scars across the land, people and trees on Tokunoshima. The war techniques including "strafing", which desecrated not only military targets but also local trees, houses and local inhabitants. The airfield on Tokunoshima was "naturalized" according to the US military, yet there was nothing "natural" about these violent actions. The deadliest and only land battles fought on Japanese soil during World War II were in the Ryūkyū archipelago, which led to immense suffering of civilians not only during the war but also in the years of the American military occupation after 1945. When Japan lost the Second World War, the Amami islands including Tokunoshima were surrendered by Japan to the US and administered by the GHQ (US military).[14] My great uncle Genichiro remembers people were so starved in this time they "could not wait for the crops to grow fully so ate them small and made a porridge from Cycad Palm seeds (Sotetsu-okayu 蘇鉄粥)."[15] This period is widely referred to as "Sotetsu-jigoku" (Cycad Hell) across islands in Ryūkyū when the seeds of the Sago Cycad Palm were ground, soaked to remove poison and eaten for survival. The palm seeds were in fact a staple food in the Amami island chain as a source of flour long before the war[16] yet the negative

13 Signs held by islanders during Kametsu new port people's assembly against the Futenma base relocation, Matsui et al.

14 Toshi, 89.

15 Toshi, 91.

16 Kuwahara and Hayward, 26–46.

FIGURE 3.9
US Troops advancing inland after landing on Okinawa
PHOTO: OKINAWA PREFECTURAL ARCHIVES

association of the flour has remained to the present. Recently delicious cookies and other baked goods are returning as the pre-war use of the flour is being revived.

Tokunoshima islanders have maintained strong anti-colonial movements for centuries. Islanders often recall resistance to the incursions of Shimazu Clan with informal trading in the face of restrictions as an important origin. Satsuma leadership imposed "The Fifteen Injunctions" over the entire Ryūkyū region from 1609 to control local authority, control foreign trade, restrict diplomacy with other nations and collect tributes.[17] This resistant

FIGURE 3.10 *Sotestu* (cycad palm) textile patterns
IMAGES COURTESY OF HONBA AMAMI OSHIMA TSUMUGI WEAVE MARKETING CO-OP

17 Chibana, 29.

FIGURE 3.11 Signature Drive by the Amami Student Association of Tokyo, April 1950, outside of East Exit, Shinjuku Station, Tokyo

spirit continued in the Postwar period when islanders traded books and other banned materials outside the view of US military between 1945–1953. During the occupation of Amami and Tokunoshima islands by the US military, a movement of the people worked constantly for reversion to Japan. One of these, the *Ryokufukai* (Green Breeze Society) group from Amami fought for return to Japan focused on politicians in Tokyo during 1949–50.[18] Furthermore, members of *Zenkoku Amami Rengo*, ("Amami Alliance," an umbrella organization that brought together different organizations) frequently held assemblies across Japan including Kobe, Miyazaki, Fukuoka, Osaka, Kagoshima, Kyoto and Tokyo to advocate for return.[19] One leader of the movement, Izumi Hōrō, a poet from Issen Town in Tokunoshima, went on a hunger strike at the peak of the resistance that led him to be known as the "Ghandhi of Japan." The principles of a people's movement outlined by Izumi, "The reversion movement is a movement of the people; a movement of the people cannot resort to violence. Our reversion movement is a peaceful one to the end. Let's go non-violently. Let's resist with non-resistance."[20] This non-resistant way of resisting continues in social movements in the islands up to today and colors the artworks I create.

18 Eldridge, 40.
19 Eldridge, 43 & 53.
20 Kagoshima Prefecture School Education Materials, 65.

After the reversion to Japan movement of the 1950s, people's movements have continued working to protect Tokunoshima island. One active member of these movements who was a critical inspiration for this artwork, Tomii Junichi, recalled the anti-nuclear electricity movement in 1974 as a youth and started his own engagement in island movements with the anti-toxic waste disposal movement in 1979.[21] For Tomii, it was after experiencing discrimination against islanders' way of life while studying in Kagoshima that he came to realize the culture, nature and customs of the island needed to be carefully protected. After organizing for the anti-base relocation in 2010, Tomii continues to participate in organizing activities for environmental protection, a nuclear-free future and supporting the world peace movement today.

> *"As for Futenma Air Force Base relocation, Henoko and Tokunoshima are impossible."*[22]

4 Relocation Plan

The US Marine Corps Air Station in Futenma is considered one of the most dangerous in the world by politicians on multiple sides due to the density of surrounding homes, schools and community facilities. A perpetual danger to the community, efforts to relocate the airbase by Japan and America have been met with steadfast resistance to the existence of the base at all. During the Okinawa Prefectural Assembly opposition is visibly made annually as well as the majority votes held in 2018 and the mordant position of Governor Tamaki Denny carrying on work by previous Governors of Okinawa. Aiming to fulfill an election promise in 2006 of "moving the base out of Okinawa Prefecture" Prime Minister Hatoyama Yuki attempted to move the Futenma Air Station out of Okinawa to the nearest prefecture of Kagoshima. The island of Tokunoshima, roughly 250 kilometers away, emerged as an alternate site with far less dense human inhabitation where dangerous fallout on schools, homes and towns could be avoided. Examining photographs of the US military show the early target of air attacks remained in the range of air force developments in the decades that ensued.

The relocation proved much harder than Hatoyama anticipated. Futenma Air Base and its proposed relocation sites of Henoko, Iejima and Tokunoshima

21 Tomii Junichi, interview with author, 30 January 2023.

22 Chinen, *Nankai Nichinichi*, 2010, 63.

FIGURE 3.12 James Jack. *Tēichigi Resistance* (*Four/Sun*), 2024. Handmade Yeddo Hawthorn ink on paper. 105 × 105 cm
PHOTO BY JING KAI

are in fact all targets remaining from the violence of the war into the present day. Immediately after the plan was announced, it was met with skepticism as all three of the island's mayors immediately called for transparency on the planned airbase relocation site, the labor and costs required as well as the supposed economic benefits of the drafted relocation. After review of the initial proposals, islanders organized public assemblies to discuss the matter where opinions were openly exchanged allowing for voices to be heard. All three mayors on the island immediately resisted the plan and worked together with local women's organizations, farming cooperatives, business associations and elderly groups to repeal the proposal. Protests were held on the island as well as in neighborhoods of Tokyo including the centrally visible area of Ginza and the federal government center of Kokkaigijidomae to show strong opposition to the relocation.

TABLE 3.3 Translation of Figure 3.12 into English

Rice	Will to Live Carefree	Resist Base	Sun & Rice Pattern (Japan & U.S.)	Resist
Treasuring children & longevity. Tokunoshima & Kyoto can change Japan	Coin	Long horizontal *kasuri*	Key	Fisheye
For the Future!!	Anger Anger	*Tsugajū*	Ancestors are Angry!	Oppose Relocation
Sea Urchin	Treasuring children & longevity. Tokunoshima & Kyoto can change Japan	*Tsubugara*	Slanted arrow wheel	Key
Absolute	*Hiba*	Firm Opposition	Flower	Agricultural Labor Union Kagoshima Prefecture

As the island of Tokunoshima was historically part of the Ryūkyū Kingdom,[23] the relocation plan was a continuation of colonial power over the archipelago by Japan. The *Kamadoo* movement of women living around the Futenma air base, and others stood in solidarity with Tokunoshima islanders in the media, at public assemblies, and in voicing their opinions to local and national politicians against the unjust burden of military bases in the islands. At the time, poet Chinen Ushi reflected upon the linked solidarity of islands in the Ryūkyū archipelago past and present,

23 "Amami Oshima was under the rule of the Ryūkyū Kingdom till the end of the 15th century, and after the Satsuma invasion of 1609, was put under the rule of the Satsuma domain." Tokunaga, 113.

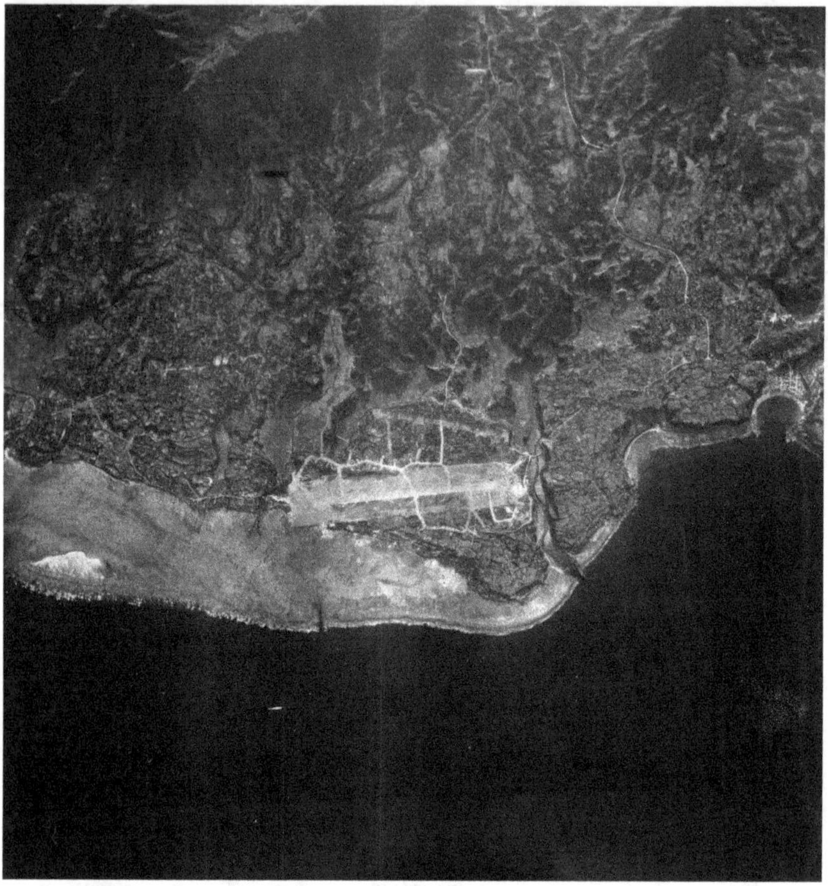

FIGURE 3.13 Tokunoshima Airfield circa 1945
PHOTO: OKINAWA PREFECTURAL ARCHIVES

When the Tokunoshima relocation plan arose, I realized once again that the Ryūkyū archipelago exists. I also understood Ryūkyū discrimination continues today. First, we want Yamato (mainland of Japan) to receive military bases and share their burden equally. Everyone already shares a consciousness of what the word discrimination means. In that period (of return to Japan) the island-wide fight led each person in turn to become stronger. There is no peace without equality. We want the mainland (of Japan) to take custody of military bases.[24]

24 Chinen in Matsui, 63.

FIGURE 3.14 Assembly against the relocation of Marine Corps Air Station Futenma in Kametsu New Port, Tokunoshima Town, Kagoshima Prefecture. 18 April 2010
(C) AFP/ISEN TOWN

Her points are clearly rooted in tradition and show the pathway to peace with justice. Standing in solidarity with other social movements resisting outside domination, islanders work beyond differences to work for the larger struggle. More than half of the island's population came to a demonstration against the base held during 2010 in Kametsu Port. In local legends islanders say there were so many people at the port that day the island tilted sideways. Grassroots support went beyond this day to include assemblies in Tokyo and Kagoshima as well as residents donating 500-yen coins to the movement in small villages across the island.[25] The tightly woven traditions of resistance were successful in thwarting the relocation plan to Tokunoshima where culture remains connected to its pasts in order to retain its identity into the future.

> *"Life is emotion."*[26]
> *"Urahoganashi ga okotte iru"* (*Our Ancestors are Upset.*)[27]

---

25 Toshi Genichiro interview with author in Todoroki, January 15, 2015.
26 Toshi, 131.
27 Matsui, 16.

FIGURE 3.15 James Jack. *Tēichigi Resistance (Two/Rice)*, 2024. Handmade Yeddo Hawthorn ink on paper. 105 × 105 cm
PHOTO BY JING KAI

5 *Kashirii* Patterns

Inside aunt Setsuko's house is the first memory I have of experiencing the deep toned threads, natural colors and rich hues of *kashirii* (カシリイ, *kasuri*/絣, ikat, dye pattern). The intricate patterns on the zabuton pillows we sat on in the living room as well as the kimono hanging in the bedroom closet where we slept were astounding. The traditions of sericulture revealed in dye, cotton and silk crafts over the past five centuries were alive in their fibers with an infinite range of earth tones within: beige, sand, kobicha, taupe, sepia, tan, burnt umber, ochre, rust, sienna, and more. Each color felt authentically connected to the land. The island tradition of submerging fibers in the earth

TABLE 3.4 Translation of Figure 3.15 into English

—	No!! Base	Tokunoshima Is Our Treasure	Water Well	Resist Base
Cycad Palm Leaves	Tree Stump	Medal	Scissors for cutting thread	River on a *Tatsugou* pattern
Against Base	Mountain path in *Tatsugō* pattern	*Hiba*	Total Resistance	Will to Protect Tokunoshima
Sun & Rice (Japan & U.S.)	Don't Play with our Islander Hearts	Rice Field	Lease Rod (Loom)	Totally Against
Carpet with *Tatsugō* pattern	*Chikoto* pattern	In Solidarity with Tokunoshima!	Sea Urchin	Carpet with Tatsugō pattern

FIGURE 3.16 Dragonfly textile patterns.
IMAGES COURTESY OF HONBA AMAMI OSHIMA TSUMUGI WEAVE MARKETING CO-OP

dorozome (泥染, mud dye) giving birth to these rich tones is often used in combination with vegetal dyes including the bark of *Tēichigi*. These combinations result in night sky shades of black, deep ochre tones of red, golden hues of yellow, diverse whites as well as colors with a purple sheen inside these dynamic prints. Each miniature pattern within the complex fabrics forms links between things in the world as a lattice of entanglements between human, vegetal and animal species. For example, some patterns within the fabrics include: *mizukumo* (flow of clouds), *uni* (sea urchins), *sotetsu* (cycad palm) and *genkotsu* (fist). These miniature patterns, known as *ka-kasuri* or "micro-mosquito" *kasuri* patterns developed in Tokunoshima and Amami islands were enhanced dramatically with the introduction of the loom for making ikat matts in the late 19th century.[28]

Intricate patterns are deeply woven into the life of islanders in Tokunoshima, and have survived historical periods of prohibition by transforming the methods of creation. *Oshima tsumugi* "hand-bound ikat" were banned by Satsuma colonizers in the islands as resources were extracted for the clan based in Kagoshima. Nature has been compounded by uneven imperial extractions of botanical resources, largely dominated by sugar cane in the islands of Ryūkyū driven by investments of the Satsuma Clan that continues in other forms of agricultural control today. Colonial incursions by Yamato cultures into the Ryūkyū archipelago over the past five centuries has not stopped the continuation of local dialects, crafts and spiritual practices.

As traditional songs were often sung while preparing materials for dyeing patterns,[29] I continue singing with Gushiken Hideaki whenever engaged in artivism today. As an artist, I work against militarism with plant and earth-based inks to care for humans and more than human species. Through anthropocentric lenses vegetal relations have been misconstrued away from plants. Yet by redirecting anthropocentric thinking as in "unlearning imperialism"[30] colorful histories of vegetal relations are awoken. For example, branches from the *fukugi* (福木, happy tree, Garcinia subelliptica) commonly planted along the coastline and in front of homes for wind protection release a bright yellow ochre color. Dyeing t-shirts with *fukugi* at the Village of Youthful Angst is a way to "create a society that does not give birth to militarism."[31] Hideaki works with powerful spirits to build bridges between indigenous

28 Tokunaga, 116.

29 Ishigaki, 2007.

30 Azoulay.

31 Gushiken Hideaki quote from *Sea Birth three*, artwork by author, 2019.

Okinawans, activists, construction workers and military personnel. Youthful weavers including auntie working in Boma with a loom create contemporary crafts in touch with traditional methods for caring. These crafts are being made within her home where threads are an intimate part of daily life. The new generation of dyers in Amami including Yukihito Kanai states "dye is attitude, not traditional" work dynamically with local earth and botanic colors in contemporary designs. Kanai's works are exhibited across Japan and gaining attention in Europe, thereby sharing peaceful ways of connection with more than human family. Threads are sustained with hopeful songs sung through hardship alongside periods of abundance sharing harvests with neighbors, friends, and family near and far.

> *Odayaka ni kurashitai ("Want to Live Peacefully.")*
> *Nonki ni kurashitai ("Want to Live Carefree.")*[32]

6 Interweaving

When heavy rains fall on Tokunoshima, red soil bleeds from farmlands across roads and flows down rivers seeping into the sea. When unsustainable crops are overplanted red dirt deposits into coral, fish and sea life with an ecological effect of diffracting sunlight, blocking oxygen absorption and slowing coral egg spawning. An island proverb says, "When the ocean becomes abundant, land crops also grow very ripe."[33] The interrelationship between sea and land has been fractured as the intrinsic value of plants, their role in barter and gift economies is threatened by colonial desire for profit. Satsuma set the prices of food products including sugar, rice and wheat during the four centuries of control in the Amami archipelago that shifted value away from crops feeding islanders towards exports. This period described as *Kokutō jigoku* ("Sugarcane Hell")[34] under Satsuma replaced crop diversity with sugarcane, a force lingering today with continued sugar cultivation subsidized by the Japan Agricultural Association. This shifted diets away from flour made with cycad palm and other vegetal ancestors that grow in abundance on the islands.

The deep rust red of *tēichigi* pulsates like our own blood, and its colors shift once in contact with oxygen like our own. Blood both nourishes our organs

32 Matsui, 2.
33 Matsuyama, 134.
34 Kamiyai, 66.

FIGURE 3.17 James Jack. *Tēichigi Resistance* (*One/Sun*), 2024. Handmade Yeddo Hawthorn ink on paper. 105 × 105 cm
PHOTO BY JING KAI

inside and scatters onto surfaces of war outside. This involves a sensory experience. Feeling is central. Phytocentric worldviews return through art created together with the voices of *tēichigi* and placards of islanders. Weaving the language of islanders in social movements along with folk patterns are combined into an artistic work nurturing spaces of interaction with ethical and tangible threads. Craft use of *tēichigi* and mud dye bath techniques were interrelated with resist methods of "binding the thread where the ikat will avoid being dyed when soaked in the dye bath."[35] These methods are the focus of this essay for their multiple meanings in craft and social movements.

35 Tokunaga, 115.

TABLE 3.5 Translation of Figure 3.17 into English

Rice Grain	Absolutely No Need for U.S. Base	Will to Live Peacefully	Resist Base Resist Base Resist Base Resist Base	Resistance
No Need U.S. Military on Tokunoshima	Strainer	Coral	Livestock Foodstuff Box	NO! Won't allow base to be built on Tokunoshima
Lattice	Flower	Coin	Dragonfly	Rice Field
Ancestors are Angry!	*Chikoto*/ *Chirashi*	Sea Urchin	No Need for Base	Resist Base
White / Miniscule Lattice	Medal	Citizens' Opinion is VETO	Rice	*Kurukomori* – Saigō pattern of Akaogi

Through resiliency, fiber craftspeople have shifted imaginatively to remain alive through periods of oppression, which gives hope even as the number of craftspeople diminishes. These patterns of resistance live in the hearts and hands of those who live not only with the threads, dyes, and fabrics but also with their entangled strength in diaspora.

In the artwork *Tēichigi Resistance*, patterns are woven together with activism that prevented relocation of an air force base onto the island in 2010 with the strong will of the islanders. Dyeing threads with rich colors cultivates knowledge of culture alive in synergistic patterns. Dyeing is a slow repetitive process of slight adjustments paired with keen observation. Leaves, bark, bacteria, humans, insects, and animals depend upon each other. One species is not centered as in human-centered thinking. The complexity of co-dependence can be felt in the process of boiling day upon day, dyeing over and over, sensitizing

FIGURE 3.18 James Jack. *Tēachigi Resistance* 2022 handmade yeddo hawthorne ink on paper (detail)
PHOTO: JING KAI

FIGURE 3.19 Odayaka ni kurashitai ("Want to live peacefully") and Nonki ni kurashitai ("Want to live carefree") all island anti-base relocation rally, Tokunoshima. 28 March 2010.
PHOTO: NANKAI NICHINICHI SHINBUN

together. Here "resistance" is woven with multiple threads: one thread arising from techniques of rice, soy and other binders that "resist" dye from penetrating fabric in order to create patterns upon being washed with water. The other stems from social reform movements of resistance against toxic impositions of colonialism, waste and militarism. Art creates patterns from colors already there by weaving them intricately with stories.

Actively working with islands as our center, Tokunoshima holds lessons in its forests and society in how to resist external impositions creating a future with us, by us and for us. Islanding the world brings them out of vulnerability and into centers that "turn the energy of the island inside out."[36] Handwritten signs speak directly to the way of life that islanders do not want to give up for militarism, money or outside control. Creativity sustains activism in the long term in colorful, engaged and imaginative ways. The interweaving of art and activism provide valuable resources for preserving people's voices and phytocentric traditions with fresh relevance today. Other vegetal dyes including *fukugi* (yellow), *gajimaru* (red), *ai* (blue), *fukugi* and *ai* (green) create a rainbow of strength. As an artist, I work towards non-violent futures for people, plants and more than us. The remnants of war are around us in many forms, knowing the past helps us dissolve these violent hierarchies and protect peaceful islander ways of life for the next generations. The question "What kind of ancestor do you want to be?"[37] weaves throughout this volume as we reflect upon human plant entanglements. Lessons from plants resist oppression and link humans with indigenous knowledge in the places where we inhabit. Devoting my hands to patterns of *kashirii* and social movements within the complexities of the present bind ancestors' love for the vegetal as part of our daily lives today.

This is a borrowed world. The time to return it will surely come.[38]

7 Weaving Generations

While hiking up Mt. Inokawa in central Tokunoshima with *Rainbow Society* members ten years ago, the intensity of islanders' care for preservation of plant species was unforgettable. In addition to common protocols such as not littering and minimizing trampling plants along the trailside, members upheld additional safeguards such as prohibiting photographs and avoiding posts on social media that might give away locations of rare flowering plants. These

36 "Shall we make "Island" a verb? As a noun, it's so vulnerable to impinging forces. Let us turn the energy of the island inside out. Let us "island" the world!" Teresia Kieuea Teaiwa "To Island" reprinted in *Indigenous Pacific Islander Eco-Literatures*, 127.

37 Hausdoerffer et al.

38 Matsuyama, 126.

FIGURE 3.20 James Jack. *Tēachigi Resistance* (*Three/Rice*), 2024. Handmade Yeddo Hawthorn ink on paper. 105 × 105 cm
PHOTO BY JING KAI

include orchid species such as *Amami Kayaran* (アマミカヤラン, Thrixspermum pygmaeum), which is not found in any other part of Japan. Efforts of citizens to document, protect and cherish rich ecological plant species such as *Solenogyne mikadoi*, *Pteris formosan* and *Asarum hatsushimae* in Tokunoshima culminated in a successful bid for UNESCO World Heritage status in 2021 along with habitats on Amami, Northern Okinawa (Yanbaru) and Iriomote Islands. The report successfully claimed "the essential value of this property consists in the fact that it contains extremely important and significant habitats for in-situ conservation of global biological diversity, providing irreplaceable habitats for terrestrial species of flora and fauna, including ninety-five endangered species on the IUCN Red List."[39] Today the UNESCO natural heritage status recognizes

39 *The Comprehensive Management Plan for Amami-Oshima Island, Tokunoshima Island, Northern Part of Okinawa Island, and Iriomote Island Nominated for Inscription on the*

TABLE 3.6 Translation of Figure 3.20 into English

Will to Live Carefree	Prevent U.S. Military Base	Angry	Square wood cups for *saké*	Resistance
Strainer, Basket for Draining	Coral Flower (Justicia carnea)	Sun	Star, Planet, Heavenly Body	No Base in Tokunoshima
No Need for Base	Will to Protect Tokunoshima	Citizens' opinion does not waver	Resist Military Base	Sea Urchin
Carpet (with *Tatsugō* Pattern)	Absolutely Intolerable	Viper Snake	RESIST U.S. Base	Dragonfly
No Need Base	Lattice	No Base In Tokunoshima	Pattern of Sun Character	No Need for Base

the value of resisting development for the preservation of more than human resources for all those who respectfully enter in the future.

Family and friends of vegetal, human and animal species on Tokunoshima show us how to live non-violently by working respectfully with treasure where we live. There are also lessons in how to care for resources that have value beyond money to protect plants, animals and people's way of life. Also, it is possible to learn methods for how to negotiate without falling into traps of polarity, through my great-uncle Genichiro who wrote,

World Heritage List. Originally adopted December 27, 2016; Revised December 21, 2018. https://whc.unesco.org/en/list/1574/documents/.

> I have always felt there are so few times when something is actually accomplished by forcefully disagreeing, from national affairs domestic and abroad, as well as those events close to me, there have always been agreements and disagreements. No matter how small a country is, those nations that preserve its seeds will always have power in all periods of time.[40]

Seeds nourished in "small" places can have a large impact on other places through networks of solidarity. Genichiro nourishes these seeds by composing songs, growing cycad palm trees and sharing stories in oral and published forms. Our relationships with other species, even those we disagree with, are important as we find ways to live with human, more than human and living beings other than us.

Devotion to the ecological heritage of Tokunoshima outlasts militaristic attempts for profit, keeping the island a peaceful and relaxed place for islanders. Resistance is rooted in the social fabric of the island and its residents. One of the lessons to share from Tokunoshima islanders with others is how to remain steadfast in one's politics amidst outside pressures to give in to demands. Daily choices of what we eat, where we live, how we travel, what materials we use, what we consume, where we dispose of things all impact our environment. These seemingly simple acts work against colonial power structures, capitalist extraction, and centralized power. Awareness of how our actions impact our descendants can be felt in the land and waters by all those who outlive us.[41] The process is ongoing and a work in progress, thus we make mistakes and fall along the way, however the resistance brings us along towards a brighter future. As one contemporary *yuta*, or shamaness of Amami Island shares, "The spirits of the earth appear. The spirits of the earth are teaching us."[42] Spirits carry us through prohibitions with hope and energy to continue blossoming for centuries to come.

The importance of passing along folk traditions to successors who can continue the crafts within the Ryukyu archipelago, is encouraged in Goya Junko's study of that which makes it possible to continue folk performance in the Yaeyama islands.[43] In this intimate study of dance, Goya notes that the idea

---

40 Toshi, 149–150.

41 "Interwoven: Our descendants will know the kind of ancestor we are by reading the lands and waters where we lived." Hausdoerffer, Ch. 4 title page.

42 Hazuki, 114–116.

43 Goya, 12.

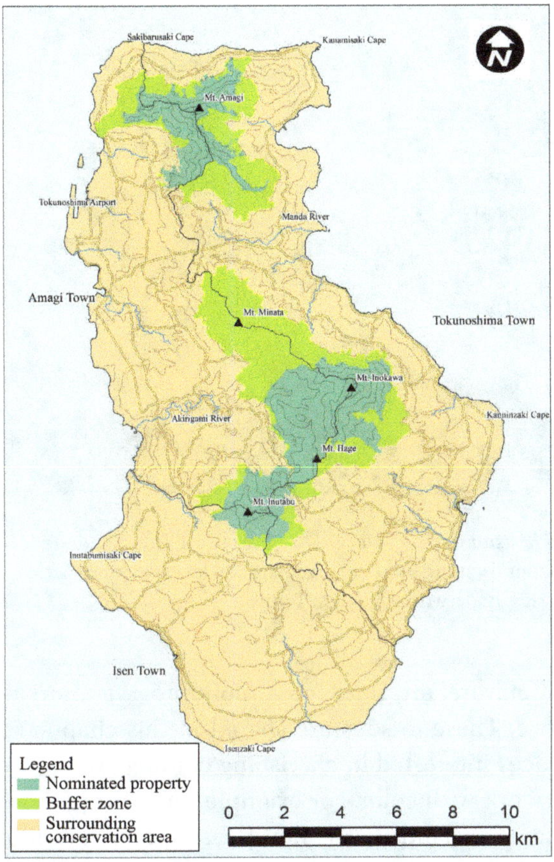

FIGURE 3.21 UNESCO world heritage areas of Tokunoshima 2021

of "tradition" in songs is a recent construction, noting their uses at the start of travel, building a new home, marriage ceremonies, pointed out that the songs have changed a lot and continue to change now.[44] How can we nurture intimate relations with our vegetal ancestors, botanic siblings and seed offspring in the everyday realities of smartphones, wi-fi and cement? Vegetal ancestors and family members teach us ways to survive in both worlds. Embrace virtual spaces as places to sing songs of dyeing, cherish the internet as a tool for practicing how to "resist with non-resistance" standing with poet Izumi today to build structures that support peaceful actions here and now.

44 Goya, 66–67.

FIGURE 3.22 James Jack. *Tēichigi Resistance* (*Four/Sun & Five/Rice*), 2024. Handmade Yeddo Hawthorn ink on paper. Two 105 × 105 cm. Installation view at Sakima Art Museum, Okinawa (Right: Artwork by Yuken Teruya)

Working with traditions of culture, art, music and more provide nourishment for long-lasting resilience. These lessons are shared in this chapter to encourage people in other places interested in cherishing culture, art, nature and heritage to continue the work with color. For example, supporting those on islands nearby Mageshima, the island that is now[45] being cornered into construction of a Japanese "self-defense" airforce landing strip with a forty-five-kilometer radius of noise pollution that impacts ecosystem of human, vegetal and animal residents. May islanders on Yakushima, Tanegashima and Nishiomote islands stand strong in their resistance to violent occupation of their air, land and sea. America continues to keep a heavy hand in island politics haphazardly preparing over the past decade to move Futenma airbase to Henoko, a site in Northern Okinawa where steadfast resistance continues daily. The impact of reclaiming the sea in the Ōura Bay with stones, dirt and cement smothering blue coral, Dugongs and rich sea life is no alternative to the site with violent impact on humans in Ginowan. Working towards peace with daily labor of love for land, connection to plants in the place where we live and respect for those who have lived here before us are radical acts of resistance.

45 Strong voices of opposition are rising against construction commencing in January 2023. https://mageshima.work/.

Like the rich colors of our vegetal friends, the abundance of "peaceful life" is coming in full bloom. In resistance to these forms of hierarchical diplomacy, island ancestors germinate and family live on. Working with vegetal ancestors is part of the decolonial work we can do towards livable futures as kin places agency of human, vegetal and other species in the pathways ahead.

Acknowledgements

This work was made possible thanks to the generosity of Furuya Chiaki, Furuya Mihoko, Toshi Genichiro, Toshi Setsuko, Toshi Morihiro, Tomii Junichi, Fujioka Yoichiro, Fujioka Heiwa, Fujioka Haruka, Fujimaki Shuichi, Kagawa Tōru, Minobe Mutsumi, Matsuoka Rieko, Sakima Michio and Uema Kanae. Artwork support provided by Ho Hui Yan Kristen, Shenali Wijesinghe, Noa Jack, and Georgette Chen Art Foundation; writing research support by Taniguchi Sotaro, Mizuno Meika and supported by JSPS KAKENHI Grant Number JP80960294.

Bibliography

"15,000 nin ga 'no' Tokunoshima de 'futenma' Hantai shūkai, (Assembly Against 'Futenma' in Tokunoshima by 15,000 people)." *Ryūkyū Shimpo*, April 19, 2010.

Alfred, Taiaiake. "On Reconciliation and Resurgence." *Rooted*, 2:1, 2022, 76–81.

Azoulay, Ariella Aïsha. *Potential History: Unlearning Imperialism*. London: Verso, 2019.

Carlile, Lonnie. "Putting Okinawa at the Center," *Asia-Pacific Journal: Japan Focus*. Course Reader #12, 2014. http://apjjf.org/data/11-8-2014_Carlile_TOC.pdf Accessed 28th December 2022.

Chibana, Megumi. *Till the Soil and Fill the Soil: Indigenous Resurgence and Everyday Practices of Farming in Okinawa*. Honolulu: University of Hawai'i Mānoa PhD Dissertation, 2018.

Chinen, Ushi, Yogi Hidetake, Toubaru Kazuhiko and Akamine Yukari. *Okinawa, datsu shokuminchi he no taido ("Quickening towards Anti-Colonialism, Okinawa")*. Tokyo: Miraisha, 2014.

Chinen, Ushi. "Kichi, hondo ni hikkitotte: Kamadou guwatachi no tsudoi", ("Accept the Burden of the Bases on the Mainland: Gathering of Kamadou Squad,") Nankai Nichinichi, 2010.

Chinen, Ushi. "Nihon koso Okinawa kara jiritsu shite (Japan be Independent of Okinawa)". *Asahi Shinbun*. August 24, 2010.

Eldridge, Robert D. *The Return of the Amami Islands: The Reversion Movement and U.S.-Japan Relations*. Lanham: Lexingon Books, 2004.

Goya, Junko. *Gakko geino: no minzokushi—sōzō sareru Yaeyama geino.* (*"The Cultural Transmission of Performing Arts in School: From the Case of Yaeyama Islands"*). Tokyo: Shinwa, 2017.

Hausdoerffer, John, Brooke Parry Hecht, Melissa K. Nelson, and Katherine Kassouf Cummings. *What Kind of Ancestor Do You Want to Be?* Chicago: University of Chicago Press, 2021.

Hayakawa, Katsumitu, Toshikazu Misao and Yoshihiro Akatsuka. "Kusukizome ni tai suru zeoraito shori no kekka" ("Effect of Zeolite Treatment on Dyeing by Plant Extracts") *SEN-I GAKKAISHI*, Vol. 50, No. 8, 1994, 364–371.

Hazuki, Mako. *Yuta: Haruka naru kamigami no shima* (*"Shamaness: Islands of Deities"*). Kagoshima: Asahi Publishers, 2014.

Hechter, Michael. *Internal Colonialism: The Celtic Fringe in British National Development, 1536–1966*. Berkeley: University of California Press, 1977.

Horton, Jessica L. "Indigenous Artists against the Anthropocene." *Art Journal*, 76:2, (2017): 48–69.

Izumi, Hōrō. *Izumi Hōrō Shishū* (*"Izumi Hōrō Poetry Anthology"*). Kagoshima: Nanpou Shinsha, 2013.

Kagoshima Prefecture School Education Materials, https://www.pref.kagoshima.jp/ba04/kyoiku-bunka/school/kyodo/documents/4990_20130604171211-1.pdf Accessed 16 March 2023.

Kamiyai, Hiroshi. *Amami motto shiritai: Gaidobukku ga kakanai Amami no futokoro* (*"Want to know more about Amami: What is not written in Guidebooks"*). Tokyo: Asahi Publishing, 1997.

Kimmerer, Robin. *Braiding Sweetgrass: Indigenous Wisdom, Scientific Knowledge and the Teachings of Plants*. Minnesota: Milkweed Editions, 2013.

Kobashigawa, Jun'ichi. "Okinawan Vegetable Dyes and Dyeing Methods," *Textiles Across the Seas. Vol. 002*. Shibuya: "Textiles Across the Seas" Executive Committee, 2004.

Kuwahara, Sueo and Philip Hayward. "Sotetsu Heritage: Cycads, sustenance and cultural landscapes in the Amami islands," *Locale: The Australasian-Pacific Journal of Regional Food Studies*, no. 2 (2012), 26–46.

Matsui, Terumi et al. *Tokunoshima no Tatakai: Futenma Hikōjo Isetsu* (*"Fight of Tokunoshima: Futenma Air Base Relocation Plan"*). August 26, 2010. Nase: Nankai Nichinichi Shinbun.

Matsuyama, Mitsuhide. *Tokunoshima no minzoku bunka* (*"Folk Culture of Tokunoshima"*). Kagoshima: Nanpo shinsha, 2009.

McCormack, Gabin and Satoko Oka Norimatsu. *Ryukyu/Okinawa, From Disposal to Resistance*. Lanham: Rowman & Littlefield, 2012.

Miyamoto, Ryūji. *Tokunoshima āto purojiekuto* (*'Tokunoshima Art Project'*), Tokunoshima Art Project Planning Committee, 2014.

Nakata, Masago. *Shima no kigitachi: Shōdoshima ueki zukan.* (*"All Trees of the Island: Shodoshima Trees and Shrubs Field Guide."*) Osaka: Olympia, 2022.

Nelson, Melissa K., Kaylena Bray and Nicola Wagenberg. "What Kind of Ancestor Do You Want to Be?," October 29, 2021, *New School at Commonweal*, podcast.

NPO Niji no kai. *Inochi tsunagaru Tokunoshima: Life Linkage-Tokunoshima.* Tokyo: Shinjusha, 2013.

Perez, Craig Santos, Kathy Jetñil-Kijiner, Leora Kava. *Indigenous Pacific Islander Eco-Literatures*, Honolulu: University of Hawai'i Press, 2022.

Ryū, Nakatame. *Nakatame nikki: bunkyū san nen ku gatsu ni jyū ichi nichi tochū yori meiji gan nen shōgatsu jyū san nichi made,* (*"Nakatame's Diary: 1863–1864"*), Tokunoshima: Tokunoshima Town Educational Committee, March 2012.

Shawanda, Amy. *Anishinaabe Motherhood: The Act of Resistance by Resurging Traditional Teachings and Pedagogies.* PhD Dissertation. Nogojiwanong (Peterborough): Trent University, October 2021.

Simpson, Leanne Betasamosake. *Islands of Decolonial Love: Stories & Songs.* Winnipeg: ARP Books, 2015.

Tokunaga, Yoshimi. "Research on the Patterns and Design of Oshima Tsumugi" *Kasuri: Textiles Across the Seas vol. 04*, Tokyo: Art Printing Co., 2007.

Toshi, Genichiro. *Hora fuki jinsei* (*"Joke Through Life"*) Kagoshima: Minami Nihon Shinbun Kaihatsu Center, 2014.

Van Horn, Gavin and Kimmer, Robin. "Kinship is a Verb," *Orion*, https://orionmagazine.org/article/kinship-is-a-verb/ Accessed March 20, 2023.

CHAPTER 4

Plant-Thinking and Critical Pedagogy in Costa Rica: The Revolutionary Praxis of Carmen Lyra

Juan Manuel Ávila Conejo

Plants, in the Western tradition, have not been considered as thinking beings. It is from this apparent contradiction that the formulation "plant-thinking" (Marder 2013) draws its generative power: from the juxtaposition of two terms that are not usually related to one another. How can we begin to define plant-thinking? Marder suggests four possible starting points: "(1) the non-cognitive, non-ideational, and non-imagistic mode of thinking proper to plants (or, 'thinking without the head'); (2) our thinking about plants; (3) how human thinking is, to some extent, de-humanized and rendered plant-like, altered by its encounter with the vegetal world; and finally, (4) the ongoing symbiotic relation between this transfigured thinking and the existence of plants" (124). To these four possible starting points, I now add a fifth one: plant-thinking, as a revolutionary way of life.

This chapter seeks to explore the concept of plant-thinking in the work of Carmen Lyra: a Costa Rican writer, communist and feminist organizer, kindergarten teacher, and anti-philosopher.[1] Lyra inhabited a world in ruins—in France. As a young woman, she experienced the human and environmental devastation brought about by World War I; and later in life encountered the same socio-ecological catastrophe in her home country of Costa Rica, as a consequence of economic crisis and plantation agribusiness. In this context, she developed a kind of thinking about pedagogy, literature, and politics based on *matas* [weeds]: the 'lowest' form of plant life and yet, the most adaptable and enduring. This chapter will argue that, in the context of the nascent plantationocene, Lyra's critical, plant-based pedagogical practice constitutes a form of revolutionary praxis that can enable new ecological ways of life age of environmental and political catastrophe.

1 Anti-philosophy "ascribes universal philosophical value to certain already-existing ordinary practices ... [it] dispenses with the heroic philosophical act and substitutes it by ascribing philosophical dignity to the practices of ordinary life." (Groys xi).

 | DOI:10.1163/9789004721777_006

1 A Short Biography of Carmen Lyra

Carmen Lyra died on May 13th 1949, in Mexico City. A few days before she passed, she wrote to her friend back home:

> Se que voy a morir, pero quiero estar por última vez en mi tierra, no quiero morir lejos de ella. Cuando no estoy en mi país me siento como mata trasplantada, de esas matas que ya sus raíces no pueden adaptarse a nuevas tierras[2] (Gonzáles, Saenz 13).
>
> I know I will die, but I want to be on my soil for the last time, I don't want to die far from it. When I'm not in my country I feel like a transplanted *mata*, one of those plants whose roots can no longer adapt to new soils.

Her last wish, to return, to die in her "soil", was denied by the counterrevolutionary regime. She died in exile, vilified as a terrorist and an enemy of the state.

Maria Isabel Carvajal Quesada, widely known by her pen-name Carmen Lyra, was born on January 15th, 1887 in San José, Costa Rica. She was the illegitimate daughter of Elena Carvajal Castro and, allegedly, a congressman by the name of Andrés Venegas. Her status as an illegitimate daughter was looked-down upon by her contemporaries; yet she was not fatherless, as her mother married Victor Quesada, a working-class man who gave her second last name. This is significant because in the Spanish naming tradition, the paternal surname always goes first and it is passed down as clan name; while the maternal surname goes second and thus disappears with marriage or death. Perhaps accidentally, her legal name refused the logic of patriarchal accumulation. She adopted the pen-name Carmen Lyra (allegedly after the streets Carmen and Lira in Santiago de Chile, where her friend and mentor Joaquín García Monge lived) and, under this name, became a popular hero.

Lyra founded the first kindergarten in Costa Rica, the Maternal School. She worked there from 1925 as teacher and director until 1933 when she was removed by the government. Her removal was politically motivated by her increasingly notorious political activism, which began in earnest in 1931 when she co-founded the Costa Rican Communist Party (later renamed *Vanguardia Popular*). She wrote fiction and children's literature during the 1920s, including her best-known work *Los Cuentos de mi Tía Panchita* (1920). After she was removed from the Maternal School, she began writing realist fiction

---

2 All translations by me.

and non-fiction political essays, fully assuming her role as *ecrivaine engage* (a French term that translates to "committed writer" or "engaged writer" in English). After the counterrevolution of 1948, she went into exile and died shortly after in Mexico.

2 Ecocritical Pedagogy

Lyra is well-known in Costa Rica for her pedagogical innovations, particularly for her adoption of Montessorian pedagogy in her work at the Maternal School (Pujol, 93). In fact, most critical literature about Lyra repeats the claim that her work was wholly or mostly based on the method developed by the Italian pedagogue. In the following section, I will complicate this narrative by showing the limits of Montessori's influence, the influence of other contemporary pedagogical theories, and Lyra's original innovations in educational theory and practice—namely what I will call a materialist, or ecocritical pedagogy.

The innovation in Lyra's pedagogy becomes clear when we examine the operation of the Maternal School: the kindergarten Lyra founded and where she worked from 1925 to 1933 alongside Margarita Castro and Luisa Gonzáles. The functioning of the Maternal School was, in some ways, influenced by the Montessori method—particularly regarding the use of games and play—yet in others the pedagogical method developed by Lyra in Costa Rica diverged significantly from the European models she learned while in France (Rubio Torres, 2).[3] It is well-established that Lyra traveled to Europe in 1920 to study pedagogy. However, there is no textual evidence of which specific authors, theories, or texts she studied while at the Sorbonne, and thus there is no way to ascertain what Lyra might have read or studied while in Europe. The strongest evidence for the Montessori influence comes from Lyra's collaborator Luisa Gonzales, who claims in the memoir *A Ras de Suelo* (1984) that they

3 Rubio Torres's article correctly identifies several innovations in Lyra's pedagogical method. However, Rubio Torres attributes these innovations to European pedagogues, namely Friedrich Froebel (German, 1782–1852) and Rosa & Carolina Agazzi (Italy, 1866–1951/1870–1945) despite the author himself noting on two separate occasions that there is limited evidence that Lyra adopted these methods from European thinkers: "No se encuentran registros del uso de los materiales diseñados por Froebel en la Escuela Maternal, especialmente los llamados *dones*, piezas de madera como pelotas colgantes o figuras geométricas tridimensionales ordenadas con rigor científico y belleza. No se evidencia, tampoco, este material didáctico en el inventario elaborado por doña Emilia Prieto ni se observa en las fotografías." (Rubio Torres 7). "Carmen Lyra y Luisa González no nombran [a las hermanas Agazzi] en los documentos inéditos o publicados" (Rubio Torres 11).

had founded the school on the "scientific methods" that they had learned from Montessori and "other modern pedagogues" (123). Thus, while it is true that Lyra followed certain aspects of the Montessori method, she also diverged from it and innovated in certain crucial aspects: the practice of the Maternal School deemphasized individual work and emphasizing collective play. However, the most important innovation was on the emphasis on the environmental aspects of pedagogy. Luisa Gonzales describes the environment of the Maternal School in the following manner:

> Por todas partes había flores aquella mañana, juguetes, columpios y subibajas, bolas y muñecas; preciosos cuadros artísticos, frisos de colores ilustrando cuentos y fábulas; la bella y buena maestra Margarita Castro tocaba al piano preciosas rondas infantiles que había escrito Carmen Lyra, llenas de gracia y poesía.
>
> That morning there were flowers everywhere, toys, swings and seesaws, balls and dolls, too; gorgeous paintings, colored friezes illustrating tales and fables; the beautiful and kind teacher Margarita Castro played the piano, beautiful children's *rondas* that Carmen Lyra had written, filled with grace and poetry [...]
>
> [...] el día de clases en que iniciaron las clases, en todas las aulas había ramos de flores y plantas, sillitas en grupos, grandes y bellísimos cuadros en las paredes, un espejo lucía en el salón central. La profesora Margarita Castro recibió a los niños mientras tocaba el piano. Las otras saludaron a cada niño invitándolo a conocer el local sin ninguna restricción: los columpios, el sube-y-baja, el mecate con nudos y otros juguetes. Carmen Lyra, la directora, los saludó y expresó: "—Chiquitos—, les dijo, quiero darles un saludo a todos ustedes. Esta escuelita es para que jueguen y aprendan muchas cosas bellas" (*A Raz de Suelo*, 123).
>
> [...] the day when classes began, in all the classrooms there were bouquets of flowers and plants, small chairs in groups, large and beautiful paintings on the walls, a mirror adorned in the central hall. The teacher Margarita Castro received the children while playing the piano. The others greeted each child, inviting them to apprehend the place without any restrictions: the swings, the seesaw, the rope with knots, and other toys. Carmen Lyra, the director, greeted them and expressed: "—Little ones—, she told them, I want to greet all of you. This little school is for you to play and learn many beautiful things" (*El primer kínder de Costa Rica*, 72).

How to explain the concern with the environment in the pedagogy that Lyra, Castro, and Gonzáles collectively developed? I suggest that the answer can be found in the contingency of the school itself. Liberalism, as the Brazilian literary critic Roberto Schwarz argues, finds itself in Latin America as an idea *fora de lugar* (Portuguese phrase that means "out of place"), a misplaced idea, due to the incongruence between the social, cultural, and economic realities of Latin America and Europe (Schwarz, 2014). In similar manner, European progressive pedagogy (be it Montessori, Froebel, or Agazzi) was quite literally out of place at the Maternal School; that is, the values that the school initially sought to embrace were at odds with the environmental reality of Latin America. The school embodied the contradiction between grandiose European ideas and the impoverished material reality of Costa Rica; between the marginal *barrios*, where the majority of the students came from, and the urbanized city center; divide between the children of proletarianized families (whose parents left the center of the country for the plantation zone) and the bourgeois families that inhabited San José; and finally, between the small, humble building of the school itself and the emphasis that the teachers gave to environmental health in the pedagogical process. As such, the Maternal School developed a multimedia approach to environmental stimulation: using music, singing, props, toys, paintings, while encouraging the children to appropriate and interact with the space on their own terms ("sin ninguna restricción"). This passage: "conocer el lugar sin ninguna restriccion" ["to know (cognize, grasp, apprehend) a place with no restrictions"] is a form of praxis against the *fora de lugar*-ness [Portuguese phrase that means "out of place"] of liberal ideology in Latin America; it is a pedagogical practice of situating (relating) the subject to their environment.

European progressive pedagogy was only nominally and briefly concerned with environmental effects on the development and education of children: this is, perhaps, explained by the fact that environmental conditions (even in the interwar period) were not as harsh in rich Western Europe as they were in post-revolutionary USSR and in Latin America. The first editions of the Montessori method touched on the importance of physical environment in the development of children. In her most famous work, *Il Metodo della Pedagogia Scientifica* (1909) the section on *Dell'Ambiente* [On the environment] that appeared in the 1909 and 1913 editions. In the sequel titled *L'autoeducazione nelle scuole elementari* (1916), there is again a section titled *Dell'Ambiente* which indicates that the transformation of the environment of the school amounts to an "opera di rigenerazione sociale" ["work of social regeneration"] (105), suggesting Montessori's concern with health was more closely related to fascist eugenics than to progressive ideas about ecological pedagogy.[4] However, the section was removed in the 1926 and later editions of the book (Montessori, 36). From this, it is fair to say that Montessori herself was at least moderately concerned with environmental health and its impact on children, but that the reception of the text de-emphasized the environmental aspects of pedagogy. In fact, as Caroli points out: "It was not until the post-war period that the child-minding aspect of their role was enriched by new psychological and educational theories about the development of language, motor activities and *relationships with parents and the surrounding environment.*" (59, emphasis added).[5] Despite the lofty idealism of pedagogues such as Delbrück and Froebel, the real historical social function of *crèches* and kindergartens in Europe was the reduction of child mortality and the emancipation of women from housework. Kindergartens, from a purely economistic perspective, were a necessary requirement for mothers to join the labour-force, and for its upkeep and enlargement. Therefore, European progressive pedagogy was not deeply concerned with the educational environment due to the relatively high standards of living; it was precisely the severe environmental conditions—what we could call, borrowing a term from Fredric Jameson, the "experience of Necessity" (87)—in marginal plantation societies that pushed Latin American critical pedagogy in the direction of environmental and ecological thought.

Having surveyed the limitations of European progressive pedagogy in terms of ecological thought, we can now move on to examine another possible influence on Lyra: Soviet revolutionary pedagogy. Although there is no direct textual evidence, there is strong circumstantial evidence that Lyra encountered Soviet pedagogy while in Paris. First, there is the fact that there were significant numbers of Soviet (mainly Russian and Ukrainian) immigrants in France during the interwar period, and they founded schools based on the 1915 progressive reform (Caroli, 2015). After returning to Costa Rica and founding of the communist party, Lyra began to openly praise Soviet pedagogy. In 1938, Lyra published an article for the journal *Trabajo*, where she recognizes the

4 On Montessori's fascist tendencies, see De Stefano *The Child is the Teacher: A life of Maria Montessori* (2022).

5 Using the distant-reading tool Google N-Gram, I was able to track mentions of *environment* in print media during the period between 1900 and 1945, in French, Spanish, and Russian language newspapers; from the historical data, we can establish that the terms *environnement* and *entorno* had a minor spike during World War I and then fall again in the interwar period (confirming Caroli's argument). In Russian, however, the result is drastically different: mentions of *окружение* begin to rise in 1915 and continue to rise significantly past the other two terms.

"new education" of "Binet, Decroly, Montessori, Clarapade, and Dewey" for its role in breaking down the "rigidity of old programs." (Lyra, 6). However, she continues, after this progressive period at the turn of the century, pedagogy underwent a process of nationalization, in which two main camps emerged: the fascist "School of Italy," led by thought of Giovanni Gentile, which she describes as purely "aristocratic" due to its exclusive, discriminatory, and hierarchical nature; and the "Soviet school," which sought to "perfect the techniques of collective labor to insure every man a dignified, free, and cultured life" (ibid). Clearly, Lyra viewed Soviet pedagogy in a positive light and shared its aim of using pedagogy in a progressive, potentially revolutionary, manner. But she goes even further, criticizing the progressive, nominally 'neutral,' liberal education as a capitulation to fascism: "between these two tendencies, one of which puts culture at the service of all and the other which only wants it for the 'best,' there is the one that takes an intermediate attitude, with its so-called 'school neutrality', which aims to hide the child true social reality." (Ibid). That being said, my argument regarding Soviet pedagogy is that there are important coincidences in method because of the significant overlap in philosophical foundations—namely Marxism and materialism—and shared global historical context; yet this should not be interpreted as suggesting that there ever was a top-down influence from the Soviet Union (or France or Italy or England) on her pedagogical practice or political thought, which would diminish the originality of her work.

In the interwar period, the newly formed Soviet Union sought to revolutionize the whole sphere of reproduction: starting with the upbringing of children, as well as the role of the women in the family and in civic life. One of the fundamental institutions in this process was the детскйи сад [lit. garden of children] and the ясли [nursery], the Soviet versions of kindergarten. Early Soviet pedagogy drew from the progressive method of Montessori and Froebel, as well as from the French model, critiquing the scholasticism and authoritarianism of Tsarist early childhood education and instead upholding the importance of freedom and play (Prokofiev, 10). However, the main ideological foundation for pedagogical practices in the early Soviet Union came directly from one of the Old Bolsheviks, Nadezhda Krupskaya, who wrote a pamphlet titled "Работница" [Woman Worker] (1889) during her exile in Siberia. In the pamphlet, Krupskaya describes the situation of working women during the Tsarist period as an intersection of capitalist and patriarchal subjugation: both as worker to capital but also "within the family, the subjugation of women to men." (Krupskaya, vi). Krupskaya goes on to denounce that working women must choose between working and raising their children, and then they must choose again between educating or feeding them. As women are forced to work, toddlers grow: "… without anyone to supervise them … They don't eat properly, get cold, go about in rags and dirty from early childhood and see their fill of drunks, debauchery, fights and much else" (14). And as soon as the child turns five, conditions become severely worse: "[Children] are drawn into cottage handicrafts between the ages of five or eight, when they are given some simple operations to carry out but work just as adults and often the same long hours. Such work is destructive of the child's organism, undermines health and blunts the child's mental capabilities. Without any moving about, without clean air in a stuffy cottage, the child grows sickly. Monotonous work from morning to night starves their intellect, does not develop it, and it becomes sluggish and stupid" (13). Already in this very early text, Krupskaya notes the relation between environment and the development of children, not just in the conditions of the space itself (stuffy air), but also in the way they interact with it (monotonous repetition).

Krupskaya then proposes in the pamphlet that the solution to the main issues that families faced during Tsarist times was the "social upbringing of children" which meant that "that the worries of supporting them will be removed from the parents and that society will provide for the child not only the means of existence but will concern itself with seeing to it that it has everything necessary for it to develop fully and in every way." (16). Krupskaya describes the ideal kindergarten in the following manner:

> Laughter and the sound of children's voices announce the presence of the kindergarten house and garden. At first glance it can seem that there is no order there but that is only how it appears. There are set programmes for their activities. They are broken up into groups and each group gets on with its business. They dig the earth, water and weed rows of plants, clean vegetables in the kitchen, wash the dishes, plane wood, glue things together, sew, draw, sing, read and play. Every game and other pursuit teaches something and the main thing is that the child is trained to be tidy, to labour, learns not to fall out with its friends and to give ground to others without caprices and tears.

For Krupskaya, the psychological atmosphere is as important as the physical environment, suggesting children should interact directly with soil and plants, and should eat the products of their own agricultural practices.

This early period of Soviet pedagogy culminated in a manual, simply titled "Ясли" ["Nurseries"] (1926) by Anna Rosenthal and Z. Shatova. The Soviet method emphasized the physical health of the children, recommending appropriate nourishment and prevention of infection diseases, which was largely successful in reducing infant mortality. Additionally, the manual stated the importance of the child's environment ["среда" lit. medium] on physical health and also for the development of the "interior world" ["внутренний мир"] of the child (Rosenthal, 51). The manual emphasized the educational environment of the child, both at the institution and at the home, arguing that parents and teachers ought to produce an environment in which the child would be able to play and interact free from adult supervision; the role of adults should be limited to creating a "favorable situation" ["благоприят обстановки"] (Rosenthal, 55). Most importantly for my argument, the manual clearly stated the importance of exposing children to nature in order to instill in them "a materialist view of the world." (Rosenthal, 55–56; Caroli, 277). It is important to note that Rosenthal and Shatova (in addition to "мировоззрение" [lit. "worldview"]) also use "воспитание," an archaic word that can be translated into English as "education" but is semantically much closer to the German *bildung* as it implies an idea of organic development of body and soul.[6] The best teacher for this materialist worldview, according to Rosenthal and Shatova, is not fantastic or religious short stories (the use of which they discourage) but "nature" itself, "which provides inexhaustible material for children's observation and research" (57).[7] This is a crucial point: the recognition of extra-human pedagogues and the de-anthropocentering of education; or to put it plainly: the recognition that we can learn important lessons from extra-human agents like animals, plants, and even the land itself. The book concludes by emphasizing the importance of the "atmosphere" of the institution, which ought to instill in children "their first habits in cooperation and an early understanding of organization in preparation for their life in a communist society" (Caroli, 277).

Krupskaya continued to play a definitive role in Soviet pedagogy as the director of the All-Russia's Congress of Pedagogy. The third edition of the congress organized in 1926 mounted a strong criticism of biologist theories of development, that is, the idea that the "psyche and personality" of children is a result of the "spontaneous development of [their] individual personality," and proposed instead that these are determined by their "conditions of life and upbringing" (Prokofiev, 12). The fourth congress explicitly declared that the "surrounding milieu determines the content and the form of preschooler's work …": much like the factory (both as form and space) conditions adult work, the form and

6 The etymology of the word is the following: вос- (*up, upwards*) + питать (*to feed*); both from Old Slavic roots.

7 "Самым лучшим педагогом в этом отношении является природа, дающая неисчерпаемый материал для детских наблюдений и исследований." (57).

space of the kindergarten instill in children a particular ideology (13).[8] From this, it is fair to say that Soviet pedagogy, based on Krupskaya's work, held a sustained interest in the educational environment and its impact on children, as well as the form of the interaction with said environment, and it emphasized these questions more than European pedagogy (although, it should be said, the real environmental conditions of kindergartens in the USSR were not better than in France and did not significantly improve until the postwar period). Krupskaya's interest in the environmental factors of pedagogy can be attributed to the materialist worldview of Soviet society: it was both a response to the real condition which she observed at the turn of the twentieth century, but it was also immediately recognized as an ideological tool of class struggle: a crucial institution in the production of a new species-being, or way of life.

Another key concern of Soviet revolutionary pedagogy was that of сказки [lit. stories, fairytales]. The concern was not with the content of the stories, but with the way they were used in the school environment. Soviet educators criticized the model of a teacher reading from a book while the children passively listened—the teachers found the "passivity" of silent listening and the stiffness of verbatim reading particularly problematic: "teachers concluded that books severed the direct tie between students and their surroundings;" and in response, revolutionary pedagogy advanced the theory of storytelling with "living" language (Kirschenbaum 69). Maria Sventitskaia, a Soviet pedagogue and a contemporary of Krupskaya, proposed a number of steps to enact this kind of *live storytelling*, including: having children illustrate the stories (through drawing and modeling), dramatization and/or storytelling by the students themselves (encouraging imaginative variation), and perhaps most importantly for my argument, Sventitskaia emphasized: "the value of conversations on natural history and the 'corner of living nature' in the Kindergarten"[9] (Sventitskaia, 72–74). Here, we can derive one of the essential features of materialist pedagogy: *to relate* children to a particular kind of nature—living nature—as opposed to the so-called human nature, that is, a nature produced by capital; or in other words, to acknowledge extra-human forms of life and learning. The term *to relate* (from the French *relation*) here carries the dual meaning of storytelling and of establishing relationship between things; a philosophical move proposed by the Martiniquan poet and philosopher Édouard Glissant.

---

8 This correct observation, however, was used to justified misguided policies that emphasized excessive academic work and deemphasized play in the 1932–1936 period.

9 "Значение бесед по природоведению и «уголок живой природы» в Детском саду." (72–74).

As Glissant explains in *Poetics of Relation* (1990), the concept of relation implied three different levels of interconnectedness: environmental, cultural, and social (146). When applied to education, we can say that an ecocritical pedagogy is one that infuses in children a sense of connectedness to extra-human life, to their language and traditional narratives, as well as to their social group; crucially, for Glissant, relation to one's social group implies the collective protection of the land or, to put it differently, Glissant's conception of the social is one which includes the land itself as member of the community. Finally, for Glissant, ecological thought necessitates the move from land to territory (lit. "territoire" 146); as this conceptual shift takes us from a merely rational/utilitarian view to a sacred understanding of the Earth. The concept of territory, and relating to the territory, also points to the materiality of earth itself, the soil [tierra], which is a crucial concept for Lyra's ecological thinking. For Glissant, as for Lyra, it is through relation (as in storytelling) that we are able to build relation (as in community) with the environment.

To summarize, Lyra's pedagogy and Soviet revolutionary pedagogy share a strong concern for the educational environment of children, this is reflected in the emphasis placed on the physical environment of the kindergarten as well as the psychological and emotional environment of the educational process itself. A key part of this environmental concern was the telling of stories in the presence of living nature: an activity that sought to instill in children an ecological (or, to use the Soviet term, materialist) sensibility that would enable them to learn from extra-human pedagogues: plants, animals, the land, the environment as a whole. These two concerns, storytelling and relation to the environment, appear throughout Lyra's works and most vividly in her best-known work: *Cuentos de mi Tía Panchita* (1920), a collection of fairytales based on African, African-American, and European tales; written as a pedagogical tool for the Maternal School.

3 The Creole Garden: Storytelling as Revolutionary Praxis

The concept of *mata* is central to Lyra's oeuvre. In the first major work on Carmen Lyra, the volume edited by Gonzáles and Sáenz, the epigraph reads: "Se que voy a morir, pero quiero estar por última vez en mi tierra, no quiero morir lejos de ella. *Cuando no estoy en mi país me siento como mata trasplantada, de esas matas que ya sus raíces no pueden adaptarse a nuevas tierras*" (emphasis added 13). Carmen Lyra writes herself as a *mata*, a Spanish word that designates small, weedy, woody plants; like peppermint or basil. The *mata* differs from the *planta* in that it is undomesticated, 'wild'; it is a form of peripheral nature, unperturbed by the world-ecology of capitalism, existing at its margins as an autarky, outside the flows of capital and biopower. Etymologically, *mata* is also related to the Portuguese *mato*, which refers to a forest, jungle, or any area filled with wild plants. Thus, *mata* carries in the Latin American imaginary the double connotation of the smallest iteration as well as the largest ensemble of wild, vegetal life. I argue that this characterization, both from herself and from her closest friends, is not a simple metaphor nor an arbitrary stylistic choice to describe the personality of the author, but instead it speaks to the epistemological mode of *plant-thinking*[10] that is at the core of Lyra's materialist pedagogy.

In order to understand this, I turn again to the work of Glissant, this time through the lens of Manthia Diawara's film *One World in Relation* (2010). In the film, Glissant speaks on *le jardin créole*, a concept that is crucial for his understanding of the intersection between relation, and ecology:

> Selon l'interprétation européenne, occidentale, l'arbre c'est la généalogie. D'ailleurs, on dit l'arbre généalogique, c'est ça. L'arbre, c'est la racine, et puis le tronc, et puis les branches au fur à mesure, des générations au

10 In the essay "Care webs and the creole garden in Manthia Diawara's *Glissant: One World in Relation*" Jeannine Murray-Román all but names the concept of plant-thinking, writing: "Beginning with the context of how enslaved persons created structures of mutual protection and distribution from the traces remaining from multiple West African cultures, Glissant gestures to the scientific knowledge employed by Africans, that of developing plant–person relationships informed by the practice of mutual protection developed among the plants themselves. For people to put into practice 'an order by which plants mutually protect each other', *they must effectively ask the plants and the earth in which they are planted whether they are disposed to grow well together: intimate interaction is the mode of learning from and with plants.*" (80–81 emphasis added).

In *Plant-Thinking: A Philosophy of Vegetal Life* (2013), philosopher Michael Marder develops the concept further, arguing that "Plant-thinking starts with the explosion of identity" (42); and that "the dispersed life of plants is a mode of being in relation to all the others, being qua being-with. Dispersed in acts of living, all creatures share something of the vegetal soul and are alive in the most basic sense insofar as they neither coincide with themselves nor remain self-contained, but are infinitely divisible below the death masks of their identities. If this is so then we have a lot to learn from plants that have mastered this way of being, which is their virtue (again, in accord with the ancient meaning of *arētē*), not a vice of insufficient self-idealization and self-universalization, as Hegel would make us believe." (51). From this vegetal being qua being-with, we can derive what Marder calls "vegetal democracy": a political form in which "the principles of inherent divisibility and participation are paramount. Inspired by the kind of sharing that marks plant-soul, which traverses all other modes of living while preserving their differences, vegetal democracy is open not only to Homo sapiens but to all species without exception." (51–52).

> fur à mesure. Donc l'arbre c'est cela qui exclut les autres. Et moi, avec quelques autres, nous pensons que cette vision-là n'est pas valable pour nous. Si on me demande de dessiner l'arbre, je dessine jamais un arbre. Je dessine une forêt. Je dessine une jungle. Instinctivement, je refuse la généalogie. Parce que sur le bateau négrier on a perdu l'arbre. L'arbre africain, le Baobab ou l'arbre sacré, l'arbre sacré sous lequel, l'arbre à palabre. Nous l'avons perdu. Parce que sur le bateau négrier, on a perdu la langue, les langues, les dieux, les objets familiers, les chants, tout. On a tout perdu. Il n'en est resté que des traces. C'est pour cela que je pense que nos littératures sont des littératures de traces. Arrivés de l'autre côté de l'Atlantique, les esclaves, ils avaient des petits jardins secrets, clandestins. Parce que, évidemment, ils étaient affamés. C'est pas les maîtres qui allaient leur donner à manger à suffisance. Et par conséquent la nuit, quand ils avaient fini de travailler, ils allaient cultiver ce qu'on appelle un jardin créole. C'est à dire, un endroit connu d'eux seulement, pour qu'on ne leur vole pas les produits. Et l'une des caractéristiques de ces jardins créoles, qu'ils avaient inventés et que nous avons perdus de vue, c'est qu'ils, dans un espace absolument réduit, ils cultivaient des dizaines d'arbres différents, d'essences différentes. Les cocos, les ignames, les oranges, les fruits à pains, les dachines, les choutchines, les patates douces, le manioc. Et ils faisaient tout ça dans un ordre tel que les plantes se protégeaient mutuellement. C'était le principe du jardin créole. Et ce principe du jardin créole, c'est le principe du rhizome. C'est pas le principe de l'arbre. C'est pas le principe de l'arbre, c'est pas le principe de l'arbre généalogique. C'est le principe de la distribution. Et nous avons perdu ça malheureusement, nous avons perdu cette science et ce savoir que les esclaves avaient. Et c'est ce jardin créole qui leur avait permis de subsister. (Diawara 13:55–17:15)

The first characteristic of the creole garden is the refusal of genealogy and the embrasure of a dynamic, dialectical, relational ecology. This observation has been confirmed by evolutionary biology, as there has been a significant shift in the last 50 years away from mechanistic models of genetics as fixed, unchanging inheritance and towards the model of epigenetics—a dynamic model that studies how the same genetic material can produce different phenotypes depending on environmental stimuli and epigenetic states (Kumar et al., 447).

The second characteristic of the creole garden is its social function as a kind of *arbre à palabre*: an east African cultural institution based on a sacred plant, usually a tree, around which the community gathers to conduct political affairs in a form of direct democracy, juridical affairs, and also serves as the site of oral-historical storytelling (N'Diaye-Correard, 41). In slave societies, in which slaves were denied all legal, political, and juridical rights, the creole garden became the space where these people could organize secretly and thus exert some limited sovereignty over themselves and their community; as such, the creole garden was the first space of self-determination within the colonial world. It is important to insist that this was not a metaphoric space but a real, physical space, in which the enslaved could speak freely, and thus organize political violence against their enslavers—the conquest of a physical space of freedom, no matter how small, is a necessary precondition to the conquest of political power, subjectivity, and dignity. But, as the name implies, the central function of the *Arbre* (and the creole garden) was storytelling—to create relation between members of the community by telling stories.

The principle of the community that the creole garden produces is "mutual protection" ["plantes se protégeaient mutuellement"]. It can be described as a place of cooperation and horizontal hierarchy. It is not a place of anarchy—there is an "order" ["ordre"] that ensures mutual protection and productivity. In a way, the creole garden can be described by the dictum: from each according to its abilities, to each according to its needs—paraphrasing Marx's formulation in the Critique of the Gotha Programme (1875). The creole garden is a place where the law of exchange value is suspended and the contingency of use value takes over. Thus, to think like a plant, means to be guided by the intimate relation between communal and individual needs.

Having introduced the concept of *plant thinking*, I must now retrace the argument back to Lyra. I have glossed over a minimal yet crucial detail: that *plantas* [plants] are different from *matas* [weeds]. The difference between a plant and a weed, of course, has no biological foundation: a plethora of research has shown the economic and empiric value of so-called weeds for indigenous, peasant, and other non-capitalist social formations.[11] However, for modern Western societies, there are significant differences between the two categories: unlike plants, weeds are undomesticated, they are autochthonous; unlike plants, which are *planted*, weeds grow without human intervention—that is, they have self-determination—unlike plants, which at best can be 'resilient' or 'resistant' (they are victims), weeds are often aggressive: spiky, woody, venomous to the touch; unlike plants, which are docile and submit to their environment, weeds often grow in inhospitable ground, like concrete or rock, and sometimes even crack and bend the environment to their will—weeds have no moral qualms about using violence, weeds are not pacifists. In terms of political economy, weeds exist outside of capitalism: they refuse to work.

11 See: Stepp (2004), Ekwealor et. all. (2019), Hassan (2020).

Or to put it more accurately, weeds resist commodification, they refuse to produce exchange value, they are unproductive. In a word—*matas* are joyous.[12] By positing herself as a *mata* and not as a plant, Lyra is refusing the object position of the slave and reclaiming—demanding—the subject position; this subject position exceeds the Hegelian dialectic in that it also refuses the position of the master, for the weed does not seek to oppress its fellow plant to gain self-determination: the weed embodies the communist subjectivity that it is able to grasp reality and History (the experience of necessity, that is to say: its environment) without succumbing to fanaticism or despair. The weed simply persists undeterred: no matter how many times it is cut down, it will grow

12 "The only worker who is productive is one who produces surplus-value for the capitalist, or in other words contributes towards the self-valorization of capital. If we may take an example from outside the sphere of material production, a school master is a productive worker when, in addition to be labouring the heads of his pupils, he works himself into the ground to enrich the owner of the school. That the latter has laid out his capital in a teaching factory, instead of a sausage factory, makes no difference to the relation. The concept of a productive worker therefore implies not merely a relation between the activity of work and its useful effect, between the worker and the product of his work, but also a specifically social relation of production, a relation with a historical origin which stamps the worker as capital's direct means of valorization. *To be a productive worker is therefore not a piece of luck, but a misfortune*." (Marx, Capital v. 1, 644).

"There is an ancient legend that might serve to illuminate the future life of communist militancy: that of Saint Francis of Assisi. Consider his work. To denounce the poverty of the multitude he adopted that common condition and discovered there the ontological power of a new society. The communist militant does the same, identifying in the common condition of the multitude its enormous wealth. Francis in opposition to nascent capitalism refused every instrumental discipline, and in opposition to the mortification of the flesh (in poverty and in the constituted order) he posed a joyous life, including all of being and nature, the animals, sister moon, brother sun, the birds of the field, the poor and exploited humans, together against the will of power and corruption. Once again in postmodernity we find ourselves in Francis's situation, posing against the misery of power the joy of being. This is a revolution that no power will control--because biopower and communism, cooperation and revolution remain together, in love, simplicity, and also innocence. *This is the irrepressible lightness and joy of being communist*." (Hardt & Negri, Empire, 413).

Lyra extolls joy as a revolutionary emotion in a speech given on Labor Day (May 1st) 1913; she addresses the children present (calling them "pequeños obreros" ["small workers"]) and asks them: "why don't we throw parties for the big workers?" She then goes on to say that Labor Day should be a joyous day, and that "en este momento, en muchas grandes ciudades, miles de ellos desfilan por las calles, con los rostros risueños y los corazones alborozados, sintiendose como los nuestros, hermanos de todos los obreros de la tierra." ["at this very moment, in many big cities, thousands of them parade through the streets, with smiling faces and joyful hearts, feeling like one of us, brothers of all the workers of the earth."] (Archive Memoria Carmen Lyra).

again in obstinacy: weeds do not give up, concede, nor negotiate; weeds are intransigent and indefatigable.

Yet, the *mata* way of life refers to a singular entity. In a letter to her friend Elias Jiménez dated August 3rd, 1916, in a highly compressed passage, Lyra considers what a collective experience of *matas* life might be like. In this letter, she writes:

> No, yo no creo que la evolución zoológica haya alcanzado la perfección. Ignoro si se alcanzará algún día. Dije sencillamente lo que mis sentidos creen percibir, lo cual puede ser un engaño, no digo que no. Dice usted: "vamos a la perfección". Sí, tal vez, pero luchando. Caminamos hacia la Armonía, escuchando los gemidos de los que resultan aplastados por los pies de los mas fuertes. Noto que a la Vida no le importan nuestros dolores ni nuestras alegrías. Para alcanzar fines que ignoramos, a veces nos hace reír y otras llorar. Quizá yo no me sé explicar y llamo "lucha por la vida" lo que en lenguaje de los que saben tiene otro nombre. (Archive Memoria Carmen Lyra)

> No, I do not believe that zoological evolution has reached perfection. I don't know if it ever will. I simply said what my senses seem to have perceived, which may deceive me, I do not deny it. You say: "we are going towards perfection". Yes, maybe, but through struggle. We walk towards Harmony, hearing the groans of those who are crushed by the feet of the strong. I notice that Life does not care about our pains or our joys. With purposes we ignore, sometimes it makes us laugh and others cry. Perhaps I don't know how to explain myself, and I call "struggle for life" what in the language of educated people might have another name.

In this letter, Lyra refers as "lucha por la vida" ["struggle for life"] to the collective experience of suffering on the way to utopia (or "Harmony"); a utopia that Lyra does not consider at all inevitable but only a possibility. Yet, the possibility of Harmony only exists inasmuch as the struggle of "those who are crushed by the feet of the strong" goes on. What Lyra calls "la Vida" ["Life itself"] encompasses both human and natural history into a great (total) collective story that subsumes all living creatures—not just humans. In a much more explicit than Jameson's *experience* of Necessity, Lyra's formulation brings us back to the (literal) ground the idea of Necessity as the actual kinetics enacted upon the materiality of the subject. History, *Life itself* for Lyra, is the boot that crushes the small; it is not simply *perceived* (as Jameson's phenomenological *experience*) nor *felt* (as Benjamin's psychological *nightmare*)—it is *endured* as a bruise, a broken bone, or a disease; and it is lived in the form of struggle. *Matas*, as vegetable beings, might not have the neurological machinery to *perceive* or *feel* oppression on a political or cultural level but as simple living creatures they undergo oppression at a much more primitive level; and it is precisely at this more primitive—and therefore, revolutionary—level that they respond to oppression. The response itself takes the form not of opposition, but of obstinacy—it is a most fundamental form of "revolutionary patience"; that is, the recognition that the "historical stratifying structures that had dominated life up to that point" cannot be dissolved within a generation, that the revolutionary impulse, if it is to be radical, must be multi-generational (Fisher, 2016).

As this point, it would be possible to connect Lyra's philosophy to Gianni Vattimo's "weak thought:" a critique of Western metaphysics as strong thought, or the thought of those in power, as a form of rationalization of what is (or, in terms of narrative a form of conservative realism) (Vattimo, Zabala, 1). Thus, weak thought as a kind of post-metaphysical philosophy "will not only surpass metaphysics but also favor its discharge, that is, *the weak who have become the vast majority of the population throughout the world*" (2, emphasis added). Weak thought derives its power from numbers—it is fundamentally democratic—on the simple premise that the so-called strong (more accurately: those in power) are few and those so-called weak (more accurately: those outside of power) are many. Vattimo and Zabala go on to propose a politics of "hermeneutic communism" not based on traditionally understood political revolution (i.e. seizing the institutional machinery of the state, which they see as a legacy of metaphysics and the tradition of scientific socialism) and instead based on interpretation and a kind of cultural revolution that is able to produce a new kind of thinking. Thus, replacing metaphysics with hermeneutics is the necessary precondition to societal change; or, to phrase it differently, the explanation of why things are the way they are must be substituted with process of systematic questioning of the status quo. In simple terms: an ideological revolution always precedes a political revolution.

Returning to Lyra, it is important to insist that although she formulates a philosophy based on the experience of small creatures, of those "crushed by the feet of the strong," at no point does she refer to this revolutionary subject as *weak*. On the contrary, as is the case with her character of Tío Conejo, Lyra's embodies the consciousness of this revolutionary world-changing thought—hermeneutics—while, *contra* Vattimo, refusing the subject position of the weak, or the Hegelian slave. For Lyra, the small life of the *mata* is the strong position, the thought of the many is the strong thought. There is no concession made. *Matas* might be "crushed by the feet of the strong," but they endure it, they are not destroyed by it. It from this way of life, of life as obstinate persistence, that the revolutionary impulse emerges.

Bibliography

Beckett, Samuel. *Nohow On: Company, Ill Seen Ill Said, and Worstward Ho*. Edited by S.E. Gontarski. New York: Grove Press, 2014.

Caroli, Dorena. *Day Nurseries & Childcare in Europe, 1800–1939*. London: Palgrave Macmillan, 2016.

Fisher, Mark. "All of This Is Temporary." Lecture, London, February 23, 2016.

González, Luisa, and Carlos Luis Sáenz. *Carmen Lyra (María Isabel Carvajal). Quién Fue y Qué Hizo?* Vol. 4. San José: Ministerio de Cultura, Juventud, y Deportes, 1977.

Hegel, G.W.F. *Phenomenology of Spirit*. Translated by J.N. Findlay. Oxford: Oxford University Press, 1977.

Jameson, Fredric. *The Political Unconscious: Narrative as a Socially Symbolic Act*. Routledge Classics. London: Routledge, 2002.

Krupskaya, Nadezhda. *Работница* [Woman Worker]. New York: Manifesto Press Co., 1901.

Kumar, Suresh, Ashok K. Singh, and Trilochan Mohapatra. "Epigenetics: History, Present Status, and Future Perspective." *Indian Journal of Genetics and Plant Breeding* 77, no. 4 (2017): 445–63.

Lemistre Pujol, Annie. *Carmen Lyra: el cuento de su vida*. 2015.

Lyra, Carmen. *Bananos y Hombres y Otros Textos*. Edited by Mijail Mondol López. Mirar al Sur. Biblioteca Literaria Centroamericana, 2020.

Lyra, Carmen. "Educación y Lucha de Clases." *Trabajo*, November 12, 1938.

Marder, Michael. "What Is Plant-Thinking?" *Klesis–Revue Philosophique* 25 (2013): 124–43.

Montessori, Maria. *Il metodo della pedagogia scientifica applicato all'educazione infantile nelle case dei bambini*. Edizione Critica. Rome: Istituto Superiore di Ricerca e Formazione dell'Opera Nazionale Montessori, 2000.

Montessori, Maria. *L'autoeducazione nelle scuole elementari; continuazione del volume: Il metodo della pedagogia scientifica applicato all' educazione infantile nelle case dei bambini*. Rome: P. Maglione & C. Strini, 1916.

N'Diaye-Correard, Geneviève. *Les mots du patrimoine: le Sénégal*. Paris: Archives contemporaines, 2006.

Prokofiev, Mikhail, ed. "Development of the Soviet Preschool Education System." *Soviet Education* 10, no. 11 (September 1, 1968): 3–24.

Pujol, Annie Lemistre. *Carmen Lyra: El Cuento de Su Vida*. 2015.

Rosenthal, Anna, and Z. Shatova. *Ясли* [Nurseries]. 1926.

Schwarz, Roberto. "Las Ideas Fuera de Lugar." *Meridional. Revista Chilena de Estudios Latinoamericanos*, no. 3 (2014): ág-183.

Sventitskaia, Maria. *Наш Детский Сад* [Our Kindergarten]. Biblioteka Vestnik Prosvesheniya. Moscow, 1924.

PART 2

Human-Plant and the Literary

CHAPTER 5

The Root of the Problem: How Trees Can Be Utilized in Climate Change Literature to Depict the Interconnectedness of Past, Present and Future

Andrea Färber

Climate change is one of the most discussed about and controversial topics in todays' world. The Sixth Assessment Report by the Intergovernmental Panel on Climate Change (IPCC) details observed and projected impacts of human-made climate change, e.g., extensive adverse losses to ecosystems, reduced food and water security, and impacts on physical and mental health of humanity (2022, 9–11). While many scientists and climate change activists alike claim that climate change is the result of human actions, there are also a significant number of people who believe that climate change is natural and that no actions need to be taken to stop it. To some extent, these climate change sceptics are correct. Climate change is a natural occurrence and has always existed. According to The Royal Society, it is estimated that the global average temperature has increased between 4 to 5 °C since the end of the last Ice Age. This rise in temperature happened over a period of approximately 7,000 years, and started 18,000 years ago (2020). Ongoing temperature analysis by NASA's Goddard Institute for Space Studies (GISS) reveal that the global average temperature since 1880 has increased by at least 1.1 °C, and the majority of this rise has occurred since 1975 with an increase of roughly 0.15 to 0.20 °C per decade (NASA 2020). This significant surge in temperature in just over a decade attest to the impact of human actions on the planet, evoked primarily by industrialization.

One central issue of climate change is that the vast timeframes over which it happens are hard to grasp for humans. It is a form of slow violence, something that unfolds slowly over a long time and space, fueled by our present actions (Nixon 2013). The shortness of a human lifespan in contrast to the amount of time it takes for temperatures to significantly impact every part of the planet makes it difficult to truly understand the rapidity at which the climate crisis unfolds. This paper aims to explore how climate change literature can tackle this issue by decentering the human and focusing on the longevity of trees instead, to demonstrate to readers the significance of caring for the

 | DOI:10.1163/9789004721777_007

planet today in order to conserve it for future generations, both human and more-than-human alike. The novels *Barkskins* (2016) by Annie Proulx and *The Overstory* (2018) by Richard Powers feature trees as main characters along with the humans. An analysis of both these works intends to achieve two things: First, it explores how trees can be utilized in climate change literature in order to disrupt an anthropocentric worldview and displace the human main characters. Second, it suggests that to make people understand the issue of time in the context of climate change, trees lend themselves as mediators because of their longevity and the impact they have on all forms of life. Through tracing the history of trees and their entanglement with human lives and cultures over centuries, as well as weaving research on trees into the stories that have the potential to permanently influence one's worldviews, these climate change novels aim to demonstrate both the impact and dependence that humanity has on these more-than-human beings and the importance of caring for them in the present to guarantee a healthy planet in the future.

1 Decentering the Human in the Anthropocene

While sceptics still deny that human actions have led to an unprecedented rapid increase of the global average temperature, others already grapple with mediating this new influence of humanity on the planet. The concept of the Anthropocene is one such attempts to give a name to the issue. The term was coined by Paul J. Crutzen and denotes an unofficial geological age marked by the dominant influence of humans on the climate and ecosystems of the planet (2002, 23). While the dominance of humanity over the planet is undeniable, there are copious amounts of worry about the significance of semiotically putting humans at the center of the planet's geological age as well. Culture and nature have been regarded as binary opposites in Western culture since Ancient Greece (Plumwood 1993, 72). The term 'nature' has been consistently shaped and reshaped by humans' needs and standards at a particular moment in time, and is thus always a projected idea of humans on the environment (Williams 1980, 82). Plumwood argues that the culture/nature dualism central to Western society is essentially what, partially unknowingly, hinders humans from conceiving themselves as a part of nature (1993, 23), and consequently halts them from taking care of the planet as necessary (1993, 71).

Scholarly efforts are trying to break this divide of culture and nature by situating the human within nature instead. One such attempt is to substitute the idea of the non-human through other terms meant to eradicate this existing dualism. The concept of the more-than-human, coined by David Abram, aims to position the human within nature as an experiencing subject rather than observing nature from a distance. Thus, humanity and its many cultures are inherently a part of the more-than-human instead of separate from it. It is the emergence and spreading of the alphabet, Abram argues, that made humans aware of their "own transcendence and timelessness relative to the fleeting world of corporeal experience" and distanced them from the more-than-human to begin with (2017, 73).

Another approach is to reject using Anthropocene to reference the present geological age, and instead term it with a focus on what needs to happen rather than the present planetary situation. Haraway criticizes the use of the term Anthropocene for its insinuations, such as destruction of environments and resource depletion of natural resources and reserves of the Earth (2016, 100). She proposes that humanity should move on from the Anthropocene to a new epoch that she has termed the Chthulucene, a period that demands humanity to both pledge itself to as well as cooperate with other inhabitants of the Earth to guarantee the thriving of all species, humanity included (2016, 101).

Climate change literature also grapples with portraying the uncertainties of the climate crisis to its readership. The more-than-human oftentimes takes on a significant role in novels to convey the necessity of taking care of the planet accordingly. Climate change plays such as Chantal Bilodeau's *Sila* (2015) and Colleen Murphy's *The Breathing Hole* (2020) situate polar bears next to the human characters and highlight their shared experiences, but also the suffering the innocent animals experience at the hands of human-made climate change. Butterflies in Barbara Kingsolver's *Flight Behavior* (2012) draw attention to the destruction of their natural environment courtesy to climate change. Besides the animal characters, there are other more-than-human characters often represented in climate change literature, ones that might be unexpected to a reader: trees.

It is undeniable that trees are a silent constant in our lives. We encounter them every day, whether in their original form or processed as wood. Yet not many people actually spend much thought on them, or even perceive them as living organisms. Trees, however, are capable of much more than what we can see. Suzanne Simard, a professor of forest ecology, has uncovered fascinating characteristics about trees after investing decades of research into the topic. These findings can change the ways in which trees are apprehended by the general public:

> Our modern societies have made the assumption that trees don't have the same capacities as humans. They don't have nurturing instincts. They don't cure one another, don't administer care. But now we know Mother Trees can truly nurture their offspring. Douglas firs, it turns out, recognize their kin and distinguish them from other families and different species.

> They communicate and send carbon, the building block of life, not just to the mycorrhizas of their kin but to other members of the community. To help keep it whole. (2022, 277)

Similarly, German forester Peter Wohlleben (2017) points out that trees are interconnected and that each and every one of them is valuable to the community (4). Furthermore, he suggests that trees can communicate through olfactory, visual, and electrical signals (12) and that new research exhibits signs of plants communicating through sound waves as well (12–13). The fungal connection that Simard argues transmits signals on the level of root systems has been termed the "wood wide web" and suggests that trees are even in contact with different species below the surface (Wohlleben 2017, 10–11).

While the idea of the "wood wide web" and the importance that Simard puts on it strengthening the entire community experiences significant pushback as being misleading (Flinn 2021, Popkin 2022), it is nevertheless proven that trees do communicate with each other and are well equipped to fight off predators—otherwise, they would not be able to survive for centuries were it not for the interference of humans. Flannery, in the foreword to Wohlleben's book, points out that the reason many of us are not able to truly understand trees is because they live on a different time scale to ours: "One of the oldest trees on Earth, a spruce in Sweden, is more than 9,500 years old. That is 115 times longer than the average human lifetime. Creatures with such a luxury of time on their hands can afford to take things at a leisurely pace" (2017, vii). To put it into perspective on a historical scale, the earliest writing that we know of, the Sumerian script, dates back to 3400 BC (Brown 2021). One of the oldest trees on Earth, then, has existed for almost twice as long as human civilization.

Both the climate and trees thus have something in common with each other: without human interference, they develop over enormous timespans that are incomprehensible to humanity. Furthermore, climate and trees are inherently interconnected with each other as well. As an outcome of climate change, trees will migrate into landscapes they typically cannot be found in, fires and severe droughts will impact forests, and consequently they will become less resilient to pests and pathogens to name just a few issues (Moore 2021). However, trees could also help be a part of the solution: "If done worldwide, natural regeneration of forests could capture up to 70 billion tons of carbon in plants and soils between now and 2050—an amount equal to around seven years of current industrial emissions" (Melillo 2021). This may not solve the climate crisis altogether, but it would certainly have a positive impact on the planet. Due to the important role that trees play when it comes to the climate crisis, it only makes sense that writers shine a spotlight on them in climate change literature in order to highlight their importance amidst the unfolding crisis.

2 *Barkskins*: Generational Interconnectedness with the More-Than-Human

Barkskins spans over three centuries and multiple generations. Charles Duke and René Sel cross the Atlantic to work as barkskins (woodcutters) in a French-Canadian settlement. Charles runs away from their master and eventually founds a lumber company in Boston, which will make his family prosperous for generations to come. René stays behind and marries the Mi'kmaw woman Mari, who already has three children. Their offspring struggle against invasive white settlers and stay in the lumber business as laborers. The lives of generations to come of these two barkskins are intertwined with each other, and deeply interconnected with the forests which dictate their living.

Notably, *Barkskins* thematizes the colonization of Native American land, and the opposing opinions that white settlers and Native Americans have on them. For white settlers, such as Duke and his descendants, wood exists to make money. They disregard trees as commodities to exploit for their own gains. René's family, on the other hand, still feels connected to nature. The need to make a living, however, forces them to still work in the lumber business and to contribute to the mindless destruction of the land.

The juxtaposition of humans and trees in this novel at first glance appears to give humans the upper hand. Industrialization allows for more trees to be felled in a shorter amount of time. The Duke enterprise expands from North America around the globe, even all the way to New Zealand. While a significant amount of the novel focuses on the destruction of the forest, it is precisely the vast timespan covered that reveals the dominance of the trees to begin with and decenters the human by the end.

At the beginning of his trading business, Charles, who at this point still carries the last name Duquet before he changed it to Duke, considers the advantages and disadvantages of timber trade in his craving for fortune: "He began to note objects made of wood: everything in the world. And it was all around him in quantities inexhaustible and prime" (73). Wood was considered an endless resource, and humans with their steadily improving ways of cutting, transporting, and processing it faster, perceived themselves as superior. Nevertheless, juxtaposing a tree with the family tree of both the Duke's and Sel's puts things into a different perspective. The book has both family trees printed at the back. This is for convenience, as over the course of 300 years, many characters are being introduced and it is hard to keep track of who is related to whom in which way. Some of these characters appear for no more than a couple of pages. Besides aiding the reader's memory in terms of family relations, the family tree offers a second valuable resource: a comparison between humans and trees that allows for the realization of how short and insignificant the human lifespan truly is. One tree can easily outlive these many generations. In fact, compared to the spruce in Sweden mentioned earlier that is thought to be approximately 9,500 years old, the multiple generations of these families would only have been on the planet for approximately 3% of the time that this tree has existed.

To draw a further comparison between these two families and trees, we can consider the family structure as that of an actual tree. Both René and Charles represent the trunk of the tree. The roots, their ancestors, are only briefly discussed. Although the roots are the most important part of a (family)tree as they nurture it, they are not being given much thought. René and Charles, on the other hand, are seen as the sturdiest part of the tree. Without their efforts, their descendants would not be able to branch out into the world—making their children, their grandchildren, and further generations the crown of the tree. The contemporary descendants resemble the top of the tree crown, the furthest removed from the family members of the past that make up the trunk. Thinking about the family as a tree in this context illustrates the temporal scale visually: Many generations have lived for the family tree to form. At least three generations are needed in order to have roots, a trunk, and at least one branch, and each of these generations only takes up a small role in the time that it takes for this family tree to grow. This comparison helps to envisage how one tree, on the other hand, can outlive at least three generations, putting the timescale into perspective.

While the novel focuses on the way that the human main characters profit from and depend on the forest to make a living, they are at the same time decentered by trees in various ways: First, while humans think that they dominate trees, which is an especially prominent assumption within the Duke family making their fortune from the lumber business, trees are repeatedly the ones to change the trajectory of the humans' lives—or cut it short at once. An example of this is the portrayal of the danger of log driving. Jinot Sel works as a log driver: "Like most woods bosses Marchand tagged Indians and *métis* for river work. Indians were the best rivermen for the long-log drives. Their quick reactions and inborn sense of balance, he said, made them agile on moving logs" (Proulx 2016, 299). He works the dangerous job at age seventeen, not out of necessity but because that is what his ancestors did as well, as his grandfather tells him: "This I know, for I, too, worked in those camps chopping trees, even as my father, Achille, did, and his father, the Frenchman René Sel, before him" (Proulx 2016, 282). While at the job, Jinot experiences the dangers of log driving, as some of the logs get stuck in the water and create a jam, and the consequences are a force of nature that humans no longer have any influence over:

> But Franceway lifted his arms and smote the bent devil a final blow. It broke, the jam quivered and immediately began to haul. Logs gushed over the edge. Jinot, in his bateau above the jam, saw a thirty-foot log rear up at the top of the releasing pile and plunge down like a falling arrow, striking Franceway square in the center of his back. Men onshore heard the crack. Franceway folded backward like a sheet of paper, his heels came past his ears and now a butcher's package of meat, he went under the grind. (Proulx 2016, 305)

This passage shows that whereas humans believe to be on top of nature—which is also suggested by the way they stand on top of the logs on the river—in the end, nature still has the upper hand. These men felled trees, but even after the death of the tree it still rises up to strike back, demonstrating that humans are at the mercy of nature after all.

A further example of trees decentering the human is the repeated mention of fires within the novel over the course of the centuries. Wildfires ravage the land and destroy sawing mills. Even more so, Jinot once again faces the dangers of the forests when one such wildfire breaks out at camp: "Rivulets of fire snaked up trees they had planned to cut. A nearby pine exploded. The noise from the approaching holocaust and the hurricane wind deafened them. Trees burst open. Nothing could exist in that massive furnace" (Proulx 2016, 386). Jinot experiences bad burns and in the aftermath of the fire, the trajectory of his life has been changed irrevocably. Within seconds to minutes, the harm that humans have done to trees is being reciprocated, reminding the reader of the fragility of human life in contrast to that of a sturdy tree.

Following the many generations of these two families over centuries in the lumber business further allows for readers to understand the interconnectedness between the past, present, and future, especially drawing on the idea of the family tree. It highlights in which ways the actions of the past can still affect the present and the future. The Duke lumber company bears great responsibility for the deforestation in North America. Many generations have profited from the exploitation and disregard of trees. Nevertheless, as *Barkskins* proves, the circle can be broken just as trees can branch out into new directions. Charley Duke, the great-great-great-grandson of Charles Duke, studies forestry at Yale and breaks the cycle of exploitation and deforestation. Instead, he starts to gain an interest in forestry conservation, breaking with traditional conventions of this family and passing the knowledge on to his child. This shows that over a long time, even centuries, people can still break out of the mold their ancestors have created and carve out a life for their own, righting the wrongs that their forbears have left behind.

The Sel family represents the other side of the coin in this scenario. When Kuntaw, Jinot's grandfather, tells Jinot about his ancestors' work with trees, he specifically points out that "So you see we have French blood in us. But René Sel did not work in lumber camps. He cut trees for himself" (Proulx 2016, 282). René's descendants are métis—they are part French on their paternal side, and they are part Mi'kmaw on their maternal side. René and Mari lived in harmony with the Earth's resources, only taking what they needed. Whereas the Duke family branches into new directions drastically different from their ancestors, the Sel family comes back full circle as their descendants take up botanical studies and forestry in order to support the environment.

For both families, their ancestors from approximately 300 years ago contributed to the present situation, whether willingly for profit or out of desperation to afford a living. Their role in deforestation has actively added to the climate change that threatens the planet by the end of the novel. The ancestors have, therefore, unknowingly unleashed a monster that only three centuries later would threaten their own descendants:

> "A great crisis is just ahead," said one scientist. "What we saw this last week—" he muttered. Sapatisia Sel thought he meant that they had been looking at human extinction. She wanted to cry out, "The forests, the trees, they can change everything!" but her voice froze in her throat. (Proulx 2016, 712)

To further highlight the interconnectedness of past, present, and future, by the end both families collaborate to combat climate change, without knowing that their ancestors have once known each other. Sapatisia, Felix, and Jeanne, descendants of René Sel, take part in a reforestation program: the Breitsprecher Tree Project, geared towards forest replanting. The project, after centuries, is all that remains of Charles Duke's trading and lumber business: "And so, over the centuries Breitsprecher-Duke had risen and fallen like a boat on the tides. Now the tide was out. [...] Only boxes of papers and several portraits remained of the old company. And a separate entity called Breitsprecher Seedlings" (Proulx 2016, 678). Their ancestors once crossed the Atlantic together to become barkskins. In the present, their descendants are trying to right the wrongs of their past. The novel expresses the interconnectedness between generations, humans, and humans in relation to the more-than-human. By situating these many generations in context with trees, the fleeting lives of humans are being emphasized in contrast to the sturdiness and longevity of trees. By ending the story in the present, threatened by the looming climate crisis, Proulx draws an intimate connection between actions in the past and effects in the future that foregrounds the threat of slow violence. By the end of the novel, the reader will have gained a deeper understanding of the correlations of actions in the present and their ramifications in the future. Whereas human lives are short, adding trees as consistent agents that experience both past and future aims to invoke a sense of responsibility toward the planet.

3 *The Overstory*: Trees as Generational and Personal Companions

Richard Powers' *The Overstory* (2018) creates intimate connections between humans and trees. All nine main characters' lives are intrinsically linked to them, some already from childhood. Like *Barkskins*, this novel too spans over generations, although the focus lies on the few decades in which the main characters explore their deep connections with the forest.

In *The Overstory*, trees are portrayed as complex beings, decentering the focus from being solely on humans. From the start, they are given a voice, "saying things, in words before words" (Powers 2018, 3). Although the trees do not talk after the initial pages, selected characters are able to hear them in their own ways: Olivia believes that they are calling out to her for help, and Dr. Patricia Westerford—a character inspired by Suzanne Simard—listens to them through her scientific discoveries. Through involvement in life-changing events, trees have influenced most of the characters' trajectories significantly. Neelay slips from the branches on his descent from an oak and becomes paralyzed from the waist down at age eleven. Douglas, working for the US Air Force in Thailand during the Vietnam War, falls out of a plane during an attack. His life is spared as he lands in a banyan fig tree, but the incident leaves him with a permanent limp. All nine main characters experience incidents like these that transform their lives in one way or another, and ultimately lead them to building lasting relationships with trees.

The novel represents the topic of the vast timespan of a tree's life in contrast to a short human life in multiple ways. First, the section titles of the book are named after the individual parts of a tree: "Roots", "Trunk", "Crown", and "Seeds". These section titles both establish a relation between the human and the more-than-human, and decenter the human at the same time.

"Roots" introduces the individual main characters, their backstories, and their—or their family's—connections to trees. These are the roots of their own lives that have shaped them; the history that has passed before their birth and the influences that have guided them to become who they are in their adulthood. Except for Ray and Dorothy, their lives are introduced separately from each other, before some of their lives start to intertwine in "Trunk".

"Trunk" is the section of the novel primarily focused on the activist work that most of the main characters take part in to try save the forests, and the backlash they receive. The attention starts to shift from their individual lives to centering around trees only. Some of the characters—Mimi, Olivia, Douglas, Nicholas, and Adam—form a group that disrupts in order to raise awareness for trees: these disruptions include tree-sittings, trespassing, and blocking roads among other things. They are met with threats and violence in their endeavor, which does not deter them. They form a strong belief that does not waver even when threatened by outside sources, solid as the trunk of a tree. Eventually, the group turns to arsonist activities to bring more attention to the issue. A failed attack causes one of their explosives to go off earlier than expected and kills Olivia. "Trunk" is the longest part within the novel, just like a trunk is typically the longest part of a tree.

"Crown" focuses on life after the incident. The characters are now living separately, trying to flee from a past that they are scared to face. Like branches in a crown, they are going their own ways over the span of decades, although sometimes they meet again, whether knowingly or not. Near the end of the novel, for example, Mimi, Neelay, and Dr. Patricia Westerford are all in the same room although they have never met before. Dr. Patricia Westerford, who struggled during "Trunk" to get approval for her work on the communication between trees, now works on a seed vault meant to preserve as many different species of seeds as possible. When she is invited to a conference to talk about the question of "What is the single best thing a person can do for a sustainable future?", she commits suicide in front of the audience by drinking from a glass of water mixed with the extract of a poisonous tree. Some of the characters get prosecuted for their past actions. Others remain free but struggle with the guilt that weighs them down. "Crown" is a representation of how the believes that they have held on to tightly during "Trunk" have had different repercussions on their futures. Each of their experiences is individual, just like each branch of a tree is individual.

"Seeds", the final and shortest part of the novel, represents flashes of their lives after Adam has been convicted over the death of Olivia. When "Trunk" represented the moments of the past that shaped the lives they were to lead, "Seeds" signals their futures yet unseen, the futures that will eventually in one way or another influence the generations yet to come. The section starts written in italics: "*Say the planet is born at midnight and it runs for one day*" (Powers 2018, 475). A comparison between the development of the planet and a twenty-four-hour day places plants up on land before ten PM, and dinosaurs have gone extinct before eleven PM. Modern day humans only show up on the clock four seconds before midnight: "*By midnight, most of the globe is converted to row crops for the care and feeding of one species. And that's when the tree of life becomes something else again. That's when the giant trunk starts to teeter*" (Powers 2018, 475). Mapping the existence of the planet on a twenty-four-hour day, a timescale that humans are intimately familiar with, particularly underscores how little time humans have actually spent on Earth. Powers rightfully raises the question of what happens when one species gains control and represses other species' chance of evolution.

The book, from a human perspective, spans over a long time that encompasses more than one generation. To structure the individual sections into parts of a tree offers a representation of how these long timespans for humans still do not signify a large time span for a single tree. It builds bridges between both humans and trees, as well as a human's past, present, and future—both those they are alive to experience, and those they do not live to see.

The Overstory also illustrates the length of times and the significance of taking care of the planet in the present to preserve it in the future throughout the text. The characters occasionally think in terms of the lifespan of a tree, rather than human time. When Adam gets sentenced to two consecutive terms of seventy years each, he is surprised by the lenience of it: "He thinks: Seventy plus seventy is nothing. *A black willow plus a wild cherry*. He was thinking oak. He was thinking Douglas-fir or yew" (Powers 2018, 471). Black willow and wild cherry both have a considerably short lifespan for trees, whereas the oak, yew, and Douglas-fir that Adam mentioned can live between 600 to 1,000 years on average.

A further example of Powers alluding to the difference of perceived time between humans and trees is towards the end of the novel, after Dorothy's husband Ray has died: "Dorothy touches the corpse's bewildered face. Already it has started to soften, even as it grows cold. "Ray?" she says. "I'll be right there." Not fast enough, at the speed of her own need. But at the speed of trees, very soon" (Powers 2018, 499). This passage once again draws the reader's attention to the fact that what the individual humans experience as long, is no more than a brief moment for trees. Ray, who has been bed-ridden for years after a brain aneurysm, spent his days looking out of the window and observing trees, finding pleasure in their existence when he no longer could keep up his own fast-paced life. By being forced to slow down life, he was finally able to connect with trees that have the luxury of taking their time.

Powers additionally builds on generational connections to trees in order to illustrate the fleetingness of human lives opposite to the planet. Introducing Nicholas Hoult in the "Roots" section, the story centers around chestnut tree on the Hoel farm. For three-quarters of a century, Hoel's ancestors have taken a picture once a month of said chestnut tree, starting with John Hoel in 1903, Nicholas' great-great-grandfather:

> When he assembles the first year's twelve black-and-white prints and riffles them with his thumb, they show precious little for his enterprise. In one instant, the tree makes leaves from nothing. In the next, it offers up everything to the thickening light. Otherwise, the branches merely endure. But farmers are patient men tried by brutal seasons, and if they weren't plagued by dreams of generation, few would keep plowing, spring after spring. John Hoel is out on his rise again on March 21, 1904, as if he, too, might have another hundred years or two to document what time hides forever in plain sight. (Powers 2018, 12)

From the start, Powers draws the reader's attention to the fact that there is much more to trees than what meets the human eye. Especially the phrase "what time hides forever in plain sight" is meant to make the reader understand the versatility of trees that we are simply not able to perceive because they develop at a much slower rate than us humans do, and because we are especially used to a fast-paced life nowadays. Through passages like these, that have characters think in tree timescales rather than in human timescales, Powers stresses the importance of looking at the bigger picture. The characters effectively detach themselves from the prevailing culture/nature dualism.

The Overstory allows for readers to realize that human life is intrinsically connected to trees in a multitude of forms, and that opposite to prevailing Western believes, humans are not superior to trees. Rather, we are inferior to them, dwelling on this planet for such a short time that we are not even able to recognize the changes in trees with our own eyes. This realization can help to also portray the climate crisis from a new perspective. Once the recognition sets in that compared to other beings on this planet, we are only here for a short while, but our descendants and other more-than-human beings around us will live in the world we have left behind, the danger of climate change as a slow violence may become more perceivable.

4 Accentuating Timescales for Awareness

As the first part of this article has pointed out, one of the biggest problems that humanity faces when it comes to the climate crisis is the fact that although it advances at an alarming pace, it is impossible to truly feel its repercussions yet due to its slow progress. This allows humans to still ignore the issue instead of counteracting it already, which will in turn affect the planet even more severely in the future.

Climate change literature with its ability to capture long periods of time, can raise awareness of this issue. Ghosh contends that time and space within a novel are typically well-defined in a narrow frame, making it an unlikely contender for portraying the climate crisis effectively (2017, 59). However, the novels discussed in this article beg to differ. They span centuries timewise, and the globe on a spatial scale. Consequently, they form nets that connect the past, present, and future, visualizing how our actions today will have an impact on the world tomorrow, even if we cannot experience it ourselves. *Barkskins* and *The Overstory* both incorporate generational aspects to raise awareness of the fact that the state of the Earth today is already an inheritance we have received. Both also take the imagery of trees and map it out on human lifespans, in the form of family trees and section titles. The longevity of trees becomes the focal point, and the human lives are arranged around it.

To conclude, climate change literature centering around trees has the unique ability to raise awareness of how us humans perceive timespans differently. Given that we only dwell on Earth for a short amount of time compared to other species and beings, the threats of the climate crisis do not appear as sever to us—rather that is out of ignorance, or because we cannot wrap our minds around the fact that our actions today can still make a difference several decades from now. By placing human characters in dialogue with trees that witness centuries if not even millennia, writers have the opportunity to show how past, present, and future are delicately interconnected with one another, disrupting the idea that the only time that matters is the present:

> "There's a Chinese saying. 'When is the best time to plant a tree? Twenty years ago.'"
>
> The Chinese engineer smiles. "Good one."
>
> "'When is the next best time? Now.'"
>
> "Ah! Okay!" The smile turns real. Until today, he has never planted anything. But Now, that next best of times, is long, and rewrites everything. (Powers 2018, 30–31)

What we did or did not do in the past affects our present, and consequently what we do in the present will affect our future. Climate change novels focusing on trees have the gift to teach this wisdom to the readers.

Bibliography

Abram, David. *The Spell of the Sensuous: Perception and Language in a More-than-Human World.* New York: Vintage Books, a division of Penguin Random House LLC, 2017.

Bilodeau, Chantal. *Sila: A Play*. Vancouver: Talonbooks, 2015.

Brown, Shelby. "Where Did Writing Come from?" *Getty*. April 27, 2021. https://www.getty.edu/news/where-did-writing-come-from/.

Crutzen, Paul J. 2002. "Geology of Mankind." *Nature* 415, no. 6867 (2002): 23. doi:10.1038/415023a.

Flannery, Tim. Foreword to *The Hidden Life of Trees: What They Feel, How They Communicate: Discoveries from a Secret World*, vii–ix. London: Willima Collins, 2017.

Flinn, Kathryn. "The Idea That Trees Talk to Cooperate Is Misleading." *Scientific American*. July 19, 2021. https://www.scientificamerican.com/article/the-idea-that-trees-talk-to-cooperate-is-misleading/.

Ghosh, Amitav. *The Great Derangement: Climate Change and the Unthinkable*. Chicago: The University of Chicago Press, 2017.

Haraway, Donna. *Staying with the Trouble: Making Kin in the Chthulucene*. Durham: Duke University Press, 2016.

IPCC. *Climate Change 2022: Impacts, Adaptation and Vulnerability*. IPCC. Contribution of Working Group II to the Sixth Assessment Report of the Intergovernmental Panel on Climate Change. 2022. https://www.ipcc.ch/report/ar6/wg2/about/how-to-cite-this-report/.

Kingsolver, Barbara. *Flight Behavior*. New York: Harper, 2012.

Melillo, Jerry. "Forests and Climate Change." *MIT Climate Portal*. October 7, 2021. https://climate.mit.edu/explainers/forests-and-climate-change.

Moore, Andrew. "5 Ways Climate Change Impacts Forests." *College of Natural Resources News*. August 31, 2021. https://cnr.ncsu.edu/news/2021/08/5-ways-climate-change-impacts-forests/.

Murphy, Colleen. *The Breathing Hole*. Toronto: Playwrights Canada Press, 2020.

NASA. "World of Change: Global Temperatures." *NASA*. January 29, 2020. https://earthobservatory.nasa.gov/world-of-change/global-temperatures Accessed: March 11, 2023.

Nixon, Rob. *Slow Violence and the Environmentalism of the Poor*. Cambridge, MA: Harvard University Press, 2013.

Plumwood, Val. *Feminism and the Mastery of Nature*. London: Routledge, 1993.

Popkin, Gabriel. "Are Trees Talking Underground? for Scientists, It's in Dispute." *The New York Times*. November 7, 2022. https://www.nytimes.com/2022/11/07/science/trees-fungi-talking.html.

Powers, Richard. *The Overstory*. London: Random House UK, 2018.

Proulx, Annie. *Barkskins*. New York: Scribner, 2016.

Royal Society. "6. Climate Is Always Changing. Why Is Climate Change of Concern Now?" *Royal Society*. March 2020. https://royalsociety.org/topics-policy/projects/climate-change-evidence-causes/question-6/.

Simard, Suzanne. *Finding the Mother Tree: Uncovering the Wisdom and Intelligence of the Forest*. London: Penguin Books, 2022.

Williams, Raymond. *Problems in Materialism and Culture: Selected Essays*. London: Verso, 1980.

Wohlleben, Peter. *The Hidden Life of Trees: What They Feel, How They Communicate: Discoveries from a Secret World*. London: William Collins, 2017.

CHAPTER 6

Multispecies Exchange Rates: Multicultural Plant Entanglement in *The Embrace of the Serpent*

Jacob Price

While the knowledge that humans and nonhumans interact with each other in ecology is undisputed at this juncture of environmental philosophy, how the terms of exchange are negotiated, carried out, and fulfilled (or not) are difficult to trace in species entanglement. When exchange takes place between two parties, each side recognizes at least one qualifiable value in the object that is transferred from one negotiating party to the other. This value can obscure the other qualities of the object for exchange and what is traded is rather the cultural, economic, or other extrinsic value imposed on the object itself. In the vast web of humans and nonhumans in the biosphere, this exchange can be involuntary yet necessary, such as trading oxygen for carbon dioxide, or decided at the edge of a shovel when breaking ground for the foundation of a new parking garage. In broader ecological thinking of what parts interchange within a whole, Timothy Morton in his book *Being Ecological* explains that each of these constituents of the whole is an endless chain of wholes and parts within itself (84). This notion complicates exchanges in that the object to be exchanged may contain multiple objects or parts within itself. When ecological exchanges take place, value is often superficial, partially understood, or reduced to a single characteristic of the whole rather than exchanges multiple parts of one whole for another series of parts of another whole. For example, when buying a Coke Zero at a store, the buyer most often purchases the soda for personal preference (the flavor) or imposed cultural values related to calories and weight (the lack of calories) but ignores the rest of what is found on the ingredient label. When humans orchestrate the exchange of nonhuman entities, the characteristics that have recognizable cultural, and subsequently anthropocentric, value are up for negotiation. By ignoring the other parts within or wholes containing the object of exchange, the bargain may even go against what the terms of the deal are, or that even the negotiating parties withhold knowledge about what exactly is being exchanged. Multicultural exchange adds another layer of complexity to negotiating a bargain, since there is no guarantee that each party understands exactly what is being offered or negotiated, even on human terms. When ecological factors are the object

 | DOI:10.1163/9789004721777_008

of trade between multiple cultures, such as the *yakruna* in Ciro Guerra's *The Embrace of the Serpent* (*El abrazo de la serpiente*), oftentimes the human protagonists make the "vegetable" character of the plant secondary, tertiary, or nonexistent in the terms of their deals. The sacred plant becomes a kind of cultural currency with fluctuating or undetermined exchange rates, where its plant characteristics are avoided or ignored by those making the exchange. Characters Theo, Manduca, Karamakate, and Evan create a cultural exchange rate of the *yakruna* to serve their own anthropocentric desire to preserve their respective cultures.

The Embrace of the Serpent presents an economic and ecological space largely outside of globalized history that the Western world desperately tries to penetrate with uneven success. At the center of this cultural exchange rate is a vegetable currency, where non-indigenous people from outside of the bioregion impose monetary value on the *yakrua* but inside Amazonia is outweighed by sacred or culture-preservative value by several of the film's characters. While the global economy assigns monetary value to rubber trees and the latex produced during the Rubber Boom, for example, the plant that protagonist Karamakate, the last known survivor of the fictional Cohiuano indigenous group, seeks is *yakruna* that is largely forgotten because it does not have an economic value during the Rubber Boom. This historical cycle of non-Amazonian economies valuing and devaluing plant life is the primary motivator behind the cultural exchange rates of the *yakruna* in *The Embrace of the Serpent*. Starting with Charles Goodyear's invention of the vulcanization process in 1839, international companies established the rubber industry in the Amazonian jungle, and by 1860 cultural exchanges between various indigenous groups and American and European businesses shaped Latin American economies and cultures (Van, 58). *The Embrace of the Serprent* begins at the turn of the 19th century, when the Rubber Boom had begun to wane, along with the economic value of Latin American latex that could not be cultivated easily in plantations versus rubber produced in Southeast Asian plantations (Smith, 19).

Yet its cultural and medicinal value carries the film forward, as German botanist Theo and American entrepreneur Evan rely on the plant to make deals with Karamakate as their guide through Amazonia. Surprising each of the main characters, the *yakruna*'s phenomenological presence reveals its own goals of exchange with(in) the minds of Karamakate, Theo, and Evan, exhibiting how nonhumans influence interspecies entanglement and exchange outside of assigned human value. The *yakruna*, as *The Embrace of the Serpent* presents the fictionalized plant, is sacred to the Cohiuano and has two primary phenomenological functions, or manifestations of its own plant agency:

objectified as currency for human cultural exchange and subjectified as the interlocutor for humans to transcend their own humanity to experience a nonhuman, plant-ified perspective of reality. Eduardo Vivieros de Castro explains this as the process of shamanism, in which an individual has the capacity to cross ontological boundaries deliberately and adopt nonhuman subjectivities (468). The *yakruna*'s phenomenological function, or subjectivity is not given by a shamanic process, but rather revealed through a ritual that that is shared across various cultures throughout the film.

At the film's beginning, the human characters and audience see the plant in the same context, defined largely by its use-value for cultural exchange. The movie starts with the initial, unsuccessful negotiation between Karamakate, and Manduca, another indigenous man from different group. The Cohiuano shaman Karamakate accuses Manduca of "submit[ting] to the whites without a fight." (3:33). Manduca explains to Karamakate that no other shamans from the region were able to cure him, and that they referred him to Karamakate. Theo's body becomes the marketplace for cultural negotiation between Manduca's and Karamakate's cultural knowledge and legacies while it reflects the bacterial exchange between his immune system and Amazonian biology. Manduca's plea is met with indifference from the Cohiuano shaman because Manduca's people, the Bará of the River, associated with white colonizers. Manduca's appearance confirms this, as he is dressed in a white button shirt and khaki pants which contrasts with Karamakate's loincloth and feathered armbands. This juxtaposition of indigenous responses to colonization not only ruptures the notion of pan-indigeneity in the Amazon, but also suggests that multiple solutions to the problem of colonization may also be in conflict: "the film offers no easy solutions for decolonisation but instead exposes a problem, namely how difficult it is to deal with processes of deculturation, acculturation, transculturation and culture preservation." (D'Argenio, 12). With no cultural common ground between the three characters in this scene, the *yakruna*, emblematic of Amazonian plant life, becomes central to bridging their interests and backgrounds. From the shared, though uneven, valuation of the *yakruna*, multicultural exchange takes place in the film, which propels the narrative forward.

Karamakate overcomes the initial impasse between the three when he notices that Theo is wearing a Cohiuano necklace, and finally agrees to help Theo in exchange for finding the Cohiuano people. Just as Argenio points out that there are no easy solutions for the problems of decolonization, likewise there are no translatable valuations of what the *yakruna* really means for each person in this scene. However, the shared "problem" among each person is the problem of cultural preservation values the *yakruna* as a multicultural vehicle of exchange. Each person brings a difficult cultural valuation of the plant, which makes the *yakruna* a kind of nexus of multicultural currency conversion. By framing the conversation through the negotiation surrounding the *yakruna*, the film positions itself outside of tropic depictions of indigenous groups: "[W]e argue that Indigenous media has been effective in an ongoing and unfinished process of decolonizing methodologies by shifting relations of power away from anthropological or other forms of hegemonic knowledge practices." (Salazar and Córdova, 134). The inclusion of various indigenous people in *The Embrace of the Serpent* with unique perspectives towards the Amazon and cultural practices disrupts the homogenization of the Amazon so that insider cultures and priorities can take center stage and actively tackle representations of decolonization, acculturation, transculturation, and cultural preservation.

Manduca participates in this multicultural transaction not as an accessory to Theo but with his own cultural investment from his position as an indigenous Bará person. During this conversation between the three, Manduca says that he believes Theo "is a wise man who has come to learn. He'll die if you don't help him!" (4:09). Manduca sees Theo as an instrument preserve his culture. His character represents the hope for cultural preservation in the Amazon through transculturalization. By adopting non-Amazonian cultural practices, Manduca leverages his position as a transcultured subject to benefit from Theo's outsider influence. Later in the film, Manduca reveals that he believes that Theo's research will advocate for all indigenous people against European self-interest: "I'm with him because we need him. He can teach the whites." (1:30:56). There is a strange contradiction of pan-indigeneity and individualized indigenous cultures at play in this transaction. For Manduca, the *yakruna* has no sacred value like it does for Karamakate but if it cures Theo, who represents scientific knowledge and an authoritative position of power among European whites, then Theo can be a vehicle for the culture preservation of generalized Amazonian indigenous groups. The *yakruna* for Manduca is an indirect means to an end, and its scientific value is more valuable than any cultural value it has.

The camera direction in this scene emphasizes each party's share of involvement in this conversation and how each convert the value of the *yakruna* to their own benefit. The camera does not tilt or pan to mimic the fluidity of the exchange, but frames Karamakate as the dominant figure in the majority of the conversation, since his cultural knowledge of the *yakruna* serves as the lynchpin to Theo's and Manduca's goals. This positioning reflects one of indigenous cinema's goals of de-emphazing a Eurocentric approach to indigenous storytelling, in which indigenous characters and perspectives drive the

FIGURE 6.1 Karamakate (left), Theo (center, back of his head), and Manduca (right) discuss the bargain between Theo and Karamakate to go to the Cohiuano and find the Yakruna. (7:30)

narrative, rather than serving as secondary devices to the film. Karmakate's position as a knowledgeable Cohiuano shaman, functions as a mediating bridge in the multicultural exchange (D'Argenio, 137). Theo may know where the *yakruna* is, but he does not know how to use it and consequently is always positioned underneath Karamakate in each take, subjugated by his sickness and ignorance. While Theo has knowledge that Karamakate wants, Karamakate ultimately does not have to render his aid to Theo. Until the last few seconds of the conversation, Manduca is behind the two, and while he does not speak, alternates between intently observing Karmakate and worriedly watching Theo.

However, after Karamakate marks Theo's face with a few circles across his left cheek and nose, the shaman moves towards Manduca to do the same. The camera cuts to the two indigenous men and as Karamakate reaches toward Manduca, his arm is slapped away. The camera frames both men positioned against each other, with Manduca becoming the dominant figure in the frame.

Manduca's defiance establishes one of the film's goals to distinguish between the large number of indigenous groups inhabiting the Amazon and their differing cultural values. By rejecting Karamakate's assistance, Manduca reinforces his insistence in Karamakate as an instrument, rather than a companion, in his quest to leverage Theo's scientific authority as an advocate for indigenous Amazonian groups. He accepts Theo's cultural position within European scientists, and though he recognizes Karamakate's ability to heal Theo, he sets a clear boundary between which cultures he is willing to rely on. While indirectly a response to the bargain at hand about finding the *yakruna*, this end to the conversation illustrates Manduca's terms of the deal.

FIGURE 6.2 Karamakate and Manduca stare each other down

Shortly after this deal is struck between Karamakate, Manduca, and Theo, *The Embrace of the Serpent* cuts to a parallel scene in which Karamakate, now much older, hears the arrival of Evan, an American scientist and entrepreneur who comes through the river and low hanging tree branches. After discovering which indigenous language Karamakate speaks, Evan inquiries about the *yakruna*, having read about it in a book Theo had published post-humous on plants in the Amazon. The ensuing conversation quickly turns into another bargain, again about finding the *yakruna*. Evan reveals what his economic intentions are with the plant: "Von Martius described it as a sacred plant that can heal diseases. That grows on the rubber and raises its level of purity" (12:49). After Karamakate responds with "Is that why you want it?" (12:52), the scene cuts without Evan's answer but continues their negotiations. Evan's interest in the *yakruna* represents the use of science to improve rubber purity and production rather than document flora in the Amazon. The evolution from Theo's goal to catalogue Amazonian flora and fauna to Evan's desire to exploit *yakruna* marks a transitory, dynamic cultural valuation of Western science and how it should apply to the Amazon. The shift in cultural and economic use-value of the *yakruna*, which symbolizes exploitable plant life from the Amazon, creates tension between Karamakate and Evan, recalling the Western economic intervention and exploitation during the Rubber Boom. As Karamakate explains later in the film, the Rubber industry's boom devasted the Cohiuano along with other indigenous groups in the region, and is one of Karamakate's principle reasons for holding prejudice against white outsiders. While the shaman understands the cultural value white Europeans and European descendants assign to science from his experience with Theo, he also has experienced trauma that rubber companies exploited indigenous groups in the name of capitalism, production, and the scientific discovery of turning natural latex into rubber from at the expense of his own culture and that of others in Amazonia.

Evan then compounds the deal by attempting to purchase information about *yakruna* from Karamakte. Evan retrieves some coca leaves and begins chewing them, prompting Karamakate to ask if Evan likes coca, and if his goal is to "take our *yakruna* too." (13:10). With this implied accusation, Karamakate hints at the history of competing ontologies towards Amazonian plant life that is a result of Western settler-colonization in the jungle. In a chapter on how different *Runa* within the Amazon rainforest understand their relationship to nonhumans, three indigenous spokespeople reiterate that:

> [i]n relations to all of nature, including the sun, moon, mountains, land, plants, and animals, we hold a cosmovision that we believe must be respected when dealing with ancestral plants. We believe that we will survive by cooperating among ourselves and sharing knowledge and the gifts of Pachamama. For us, the plants and animals are all a part of the connections between the *Runa* (indigenous people) and Pachamama. Our cosmovision is integral, guiding all decisions and strategies. For this reason, we do not speak of "conservation" but of coexistence of all life forms. (Fueres, Flores, Ramos, 108)

Karamakate's concern with the *yakruna* is more than conservation for the sake of biodiversity, a strategy deeply embedded within a Western scientific framework, but rather emphasizes the kidnapping-like nature of "tak[ing]" plants out of place and commodifying them. Reacting to the accusatory tone, Evan offers "a lot of money" and the camera cuts to reveal that he holds out two wrinkled one-dollar bills. The humor is not lost on the audience or Karamakate, as he rejects the valueless offer, joking that he will not take the cash because "it tastes bad" (13:45). Cultural attitudes towards money as currency exchange in these negotiations, as the scene parallels this moment with the bargain accepted before between Karamakate, Theo, and Manduca. Theo, coming to Karamakate with self-interest overcome his diseases, has the trust and advocacy of Manduca who vouches for him. Evan, however, does not bring any genuine or selfless understanding of indigenous cultural to the table, and there is no indigenous person to vouch for him.

The absence of a character like Manduca notably comments on the Evan's character. He arrives at Karamakate speaking several indigenous languages, but not having made any identifiable relationship with indigenous people. Whether the lack of a character like Manduca who represents an indigenous group that negotiates their culture and place in the Amazon with white outsiders signifies that this type of indigenous group is extinct, fully assimilated to the settler-colonizing culture, or simply missing does not matter. Evan, unlike Theo, did not seek out any advocacy within Amazonian cultures, but initially would rather find the *yakruna* on his own terms, which reflects economic and entrepreneurial potential rather than local cultural value. His attempt to buy information from Karamakate with money reflects how he approaches multicultural negotiation. Representative of sly American business practices, Evan's attempt to trick Karamakate, assumes that the Cohiuano shaman has assigned economic value to American money at least to the extent that Evan can exploit him. However, Karamakate's rejection of money stagnates the cultural negotiations over the *yakruna*. For him, the *yakruna* will never have economic value, and he would rather protect its inscription into a global capitalistic market than sell it.

Appearing defeated, Evan hesitates before returning to the bargaining table from a different angle. Commenting that "[he's] never dreamt" (13:51), Evan appeals to the local cultural value assigned to sacred plants with hallucinogenic properties. Despite the rocky negotiations between the two up until this point, Karamakate's interest is visible on his face, as he recounts a dream he had about another white man who also needed to learn to dream. The proceeding monologue recalls his experience with Theo. This memory, combined with Evan's recognition of the local cultural value of dreams and hallucinogenic plants, has more value to Karamakate than anything of economic potential that Evan can offer. Remembrance becomes the focal point of discussion in the remainder of the scene, as Evan offers to share coca with Karamakate. In a somber tone, Karamakate admits that not only he had forgotten how to make *mambe*, a roasted and crushed powder made from coca leaves, but that he "forgot *mamba* existed" (15:41). Evan picks up Karamakate's tray of coca leaves, and proceeds to make *mambe* for them both to consume. This exchange of knowledge between people from differing cultures undermines any preconception of notion as shamanic knowledge being exclusive to indigenous people and foreshadows later discussions about plant use-value in the film.

Despite his cultural position regarding the economization of Amazonian plant life, Evan's ability to make *mambe* expands his relationship to plants beyond Western-defined economic use-value. The inversion of expectations in this scene challenges exotifying tendencies to view indigenous identifies and cultural knowledge and practices as exclusively other, static, and/or inaccessible to outsider groups. While Karamakate sees through his dubious business ethics which contrast sharply with Karamakate's judgment of money, Karamakate also recognizes, and believes, that Evan is a man who is devoted to plants beyond typical Western valuing of plant life because he knows how make *mambe*. Karamakate does not appear bothered or angry that Evan knows how to make *mambe* though he has forgotten. On the contrary, Karamakate appears to be indifferent, as he does not watch Evan make the *mambe*, but accepts it calmly when offered to him. After sharing it together, Evan asks again if Karamakate will accompany him to find the *yakruna* and this time the shaman agrees.

In this second conversation, the preservation of cultural knowledge takes on a new form in the discussion and subsequent consumption of *mambe*, as the shared factor between divergent cultures. Unlike Manduca's defense of Theo, the preservation of cultural knowledge in this negotiation between Evan and Karamakate is not entirely theoretical or dependent on if Theo is successful or not. The newer debate, occurring many years after Theo, Karamakate, and Manduca first meet, reveals that Amazonian indigenous cultural knowledge has been preserved in some capacity. Evan even mentions the Bará, the indigenous group with which Manduca identifies, in his initial encounter with Karamakate, illustrating that perhaps Manduca's wish had been fulfilled. But Evan, as a white American man and outsider to Amazonia, has preserved the indigenous recipe for *mambe*, when Karamakate, a Cohiuano shaman, has lost that knowledge. In the final scene of the film, Karamakate will reciprocate the transferal of indigenous cultural knowledge for preservation through the vegetable agency of the *yakruna*.

When both men find the last known *yakruna* plant growing on top of a mountain, Evan panics when he sees Karamakate pulling the petals off the flower. For the final time, their cultural values and their desire for the *yakruna* conflict. Evan reveals, in the film's only reference to World War II, that a war is going on and that Americans need a high-purity rubber reminding Karamakate that the *yakruna* can fulfill that need. Angered at Evan, Karamakate feels betrayed and asks "You want to make weapons with *yakruna*? You want to turn it into death?" (1:51:41), recalling Karamakate's previous accusation that Evan wants to take the *yakruna*. This back-and-forth is framed with a shot of Evan walking across the mountaintop, surveying the horizon and expanse of the Amazon stretching out towards the rest of the world, and Karamakate next to the flowering *yakruna*.

The shot framing Evan between two dead branches, both well inside the frame of the take, suggests Evan's narrow-minded valuation of the *yakruna* and the interior world of the Amazon. He looks out towards the open world, the cross-continental conflict, where the Axis and Allies wage war against each other, hopeful that what he is doing in the Amazon turns the tide in his favor. Karamakate is positioned next to the *yakruna* with a backdrop of the lush

FIGURE 6.3 Evan stares out away from Karamakate and the yakruna

FIGURE 6.4 Karamakate, indignant, asks if Evan plans to use the yakruna for the fabrications of weapons and death

Amazonian Forest, framing Karamakate as an insider to Amazonian ontology and geography. As Holly May Treadwell explains, when Karamakate interrogates Evan to know if he wants *yakruna* to make weapons and turn it into death, the Cohiuano shaman reveals his awareness of capitalist destruction and exploitation within the context of the rubber boom (64). Evan's Western ideology looks away from the sources of the resources that his country needs and repeats the historical notion of exploitative resource extraction fueled by capitalism in Latin America, while Karamakate is decidedly couched within the Amazonian ontological framework. This separation marks the necessity for Evan and Karamakate to negotiate their cultural differences despite both of them depending on each other to find the *yakruna*. Ironically, both Karamakate and Evan desire the same thing: the preservation of their ideologies and cultures, although their methodology and ontological perception of nature varies in how they hope to achievement cultural preservation. When Evan sees Karamakate tear the petals off the *yakruna* flower, he threatens to kill Karamakate, illustrating how his culture's willingness to employ violence to preserve his culture at the cost of Karamakate's cultural knowledge and life. Evan eventually backs down allowing his individual morals to win over the violent capitalist justification to kill anyone opposed to its expansion and control over plants and nature broadly. This change and realization of his own actions create an opportunity for Evan to become open-minded to other potential use-values of the *yakruna*.

Yielding to Karamakate, Evan sits bitterly away from the shaman as the other prepares *caapi*, the substance made from the *yakruna* that will allow Evan to dream. After the *caapi* is ready, both sit for one last discussion about the *yakruna*. Karamakate reveals that he "wasn't meant to teach his people. [He] was meant to teach [Evan]" (1:55:00). Karamakate believes that the *yakruna* still can be used to preserve his culture and knowledge, but not through the means he had envisioned. This realization marks a profound shift in how Karamakate understands his own culture, and how he can extend his own entanglement with the Amazon to people outside his heritage. In this scene, Karamakate resembles Manduca from the first scene of the film, who has placed his trust in Theo as someone from outside of the Amazon who can adopt Cohiuano knowledge and cultural practices. Both instances recognize the precarious situation of indigenous ontologies in globalized politics, where foreign capitalist industry and war, regardless of indigenous peoples' awareness of their mechanisms, impact Amazonian ecologies. By having Evan consume the *caapi* and then dream, the shaman obligates Evan to retain and disperse Cohiuano culture. In this transaction, the *yakruna* becomes key in entangling culture from one person to another. Ana Maria Mutis refers to this multicultural exchange as one that implicates nature beyond anthropocentric valuations of nature:

> En contraste, *El abrazo* traza un nuevo orden en el que la escritura y la documentación histórica—así como la ficción que estas inspiran—proviene del intercambio de diferentes culturas, de la colaboración de indígenas y exploradores con la participación activa de la naturaleza. El mensaje de la película parece ser que solamente producido de esta manera el discurso puede irradiar conocimiento y despertar conciencias y así cambiar el destino de la región. (Mutis, 38)

> In contrast, *Embrace* traces a new order in which historical documentation and writing—such as the fiction these inspire—comes from the exchange of different cultures, from the collaboration of indigenous people and explorers with the active participation of nature. The film's message seems to only come about this way, the message radiates knowledge and awakens awareness, and thus changing the region's destiny. (Translation mine)

As Mutis argues, *The Embrace of the Serpent* eventually allows for the active participation of nature rather than existing as a backdrop to human interests or conversations about nature or multicultural exchange. The consumption of the sacred plant in the final scene, where its hallucinogenic properties take hold of Evan, is the first instance in the film where the *yakruna* becomes more than a desired object, but rather a subject that exhibits agency by actively engaging Evan in Amazonian ontology, rather than him learning about it second-hand from Karamakate.

Contrasting the methodical pacing present in the up-to-now black and white film, the camera cuts to several quick cuts of aerial shots of the Amazon River snaking through the jungle before bursting into kaleidoscopic colored images.

These images, taken at face-value and according to the level of engagement that the viewer cedes, force Evan as well as the viewer to recall what Karamakate has taught about *yakruna* and his general ontological vision of the Amazon bioregion throughout the film to comprehend what is on screen. The vision does not necessarily need to make sense, in the Western sense of logic. Enrique Bernales Albites connects the origin myth of the Amazon with the dreamer's necessity to listen and identify, rather than converse with the mythological serpent: "As Guerra's film depicts, dreaming in the Ayahuasca ritual enables contact with the secret powers that share messages, their songs, the love for their world. Karamakate admitted to Evan in his language: "The world speaks. I can only listen. Listen to the song of your ancestors, not only with your ears"" (Bernales Albites, 209). For Evan, despite his inability to respond to his vision while experiencing it, listening/seeing is a cultural exchange because in the *caapi* ritual, he finds validation of Cohiuano ontology through the plant's psychological mediation between the Serpent and Evan. Karamakate's advice to Evan to "listen to the song of your ancestors" implicates Evan's history within Amazonian ontology, creating a window through which Evan can connect with the vision more fully and personally through his entanglement with the *yakruna*.

FIGURE 6.5 Bold amber and orange curves with blue and red flecks that flow downward on the left and upward on the right

The *yakruna* facilitates this exchange of ontological authentication for Evan's responsibility to preserve Karamakate's cultural knowledge. Eduardo Kohn explains how multispecies negotiation takes place through dreaming, which hallucinogenic plants like the yakruna create: "Because dreaming is a privileged mode of communication through which, via souls, contact among radically different kinds of beings becomes possible, it is an important site for this negotiation" (Kohn, 140). While Kohn's book focuses on the *Runa* in Amazonian Ecuador, the film showcases several conversations about dreaming in comparable ways. The dream, or vision, that Evan experiences hosts communication between Evan and the Serpent. The *yakruna*, as a communicative medium, manifests itself as the interlocutor between both souls in the dream. The dream cannot be understated as a site of negotiation, through which Amazonian ontological nonhuman beings manifest themselves and their desires as, what Kohn calls, selves. Karamakate recognizes that making Evan dream is fundamental to preserve Cohiuano cultural knowledge because Evan's human soul will communicate with souls embodied in other species.

The multispecies character of this communication opens an ever-spanning range of possible nonhuman agencies to engage with Evan. Kohn refers to selves as both origin and product of an interpretive process (75), utilizing Charles Peirce's framework of semiotics to explain how multispecies communication operates in Amazonian ontology. When selves manifest themselves, they produce signs that are interpreted by other selves. These interactions and interpretations, producers, and receivers of signs, form the basis for multispecies exchange. In *The Embrace of the Serpent*, the Serpent is the self of origin that transmits the signs and the *yakruna* is the site of negotiation and self that in turn receives those signs and rebroadcasts them to Evan's minds as signs. The *yakruna* becomes a self that expresses signs to Evan while he is dreaming. Evan receives the *yakruna*'s signs which are plant-based units of communication before he can interpret them. This process is a kind of seamless translation of language spoken in multispecies semiotics, where the embodies of each speaker, or self, does not impede communication. However, the degree to which each self comprehends and employs learned knowledge across dreamed communication will vary depending on what the self chooses to do next.

Karamakate risks the future existence of Cohiuano cultural and knowledge, trusting that the *yakruna*'s communicative properties will effectively teach Evan about dreaming and convince him that the *yakruna* has value beyond its ability to produced high-grade rubber. When Evan awakes after his vision, he finds himself alone on the mountaintop with the *yakruna*-induced vision of the reality of Cohiuano shamanism and ontology. He is fundamentally changed and charged with the responsibility to know what the *yakruna* has taught him. The final take illustrates this newfound recognition, as after several cuts of Evan walking through parts of the jungle, he strides alongside a riverbank, only to stop when several butterflies whirlwind around him.

Characteristically similar to the flickering blue and red specs moving in tandem with the bold river images of the *caapi* vision, Evan finds himself in tune with his surroundings. This shot of Evan surrounded by the butterflies is parallel to an earlier image in the film in which the butterflies circle Karamakate. While scholar Joey Whitfield asserts that the shamanic vision is reduced to the unfortunate solving of a Westerner's depression (190), the comparison between Karamakate and Evan being surrounding by the circled butterflies suggests that Evan at least recognizes that, even without fully understanding the vision, that it means something beyond his ontological self. Timothy

FIGURE 6.6 Bright white insects swirl around Evan, who stands motionless next to the riverbank

Morton explains this as the concept of attunement, which is the act of becoming more consciously aware of being ecological (60). This scene depicts Evan as a human that is more fully comprehends what his being or self is within the greater multispecies entanglement of the Amazon. By ending the film with this imagery connecting Evan and Karamakate as humans attuned to the nonhumans around them, Evan has moved beyond his initial concern about his inability to dream, with dreaming becoming a gateway to attunement. Rather, Evan is shown in stoic shots of him walking through the jungle, and this attunement recalls Karamakate's seeming indifference and bluntness throughout the film. The insects themselves represent how his surroundings are also in tune with him. The butterflies in this shot confirm this attunement. Their swirling behavior represents their attunement with Evan. The depiction of the flowing orange, red, and yellow lines in the vision also mimic the notion of attunement as multiple species recognize how their presence has ecological consequences, intended or not, for other Amazonian beings. Evan's attunement places him within the already attuned Cohiuano ontology.

While Morton's philosophizing regarding attunement explains how Evan comprehends Cohiuano ontology, the way the movie translates that middle ground of cultural/ontological notions of ecology for the audience are not as clean cut. The film's vision scene invites a strong moral reading, having shown not only the shamanic vision to Evan but to the audience as well, supposes that all viewers have this responsibility to preserve indigenous culture through memory and acute awareness of disappearing indigenous cultures and ontologies. *Embrace of the Serpent* does not shy away from this messaging, which entangles how the film handles the convergence of multiple cultures through negotiation: "[The filmmakers] recourse to the universal value of all knowledge demonstrates its explicit attempt to perform a redemptive act vis-à-vis Western sources. *El abrazo*'s pedagogical discourse therefore employs framing devices that are consonant with the knowledge-politics the film endorses" (Gleghorn, 42). How the audience will interpret that messaging or act upon it remains their decision as a meta-narrative way for the film to reiterate the need to listen or interpret multispecies communication. Evan may have experienced ecological entanglement via hallucinogenic attunement after experiencing the shamanic vision, but viewers may simply understand the vision as film rather than an invitation to experience Amazonian ontology.

The nonhuman exchange at the film's finale, represented in the shamanic vision, allows the *yakruna* to transcend all the multicultural exchanges present in the movie into a multispecies exchange. Cultural knowledge and the desire to promote or preserve cultural knowledge designate the primary motives of each of the human characters in the film as they negotiate their own cultural value of the *yakruna* to achieve their own selfish goals. *The Embrace of the Serpent* turns the notion of the plant economic value on its head by allowing plants to no longer function as an object with an anthropocentric value, but rather asserting its own desire as an agent to connect humans with the Serpent. Karamakate knows this, as his ontology allows for the agency of nonhumans to exhibit desire and influence on humans. Evan represents a Western ontological categorization of plants evolved from the Enlightenment as objectified lifeforms: living, but without desire or agency, as machine-like entities that humanity values according to their anthropocentric use-value. With Evan's ontology colliding with the *yakruna*, he becomes an object in the Western sense through the *caapi*-induced vision. This inversion transforms Evan into a type of currency of multicultural and multispecies exchange as he becomes the vessel to transmit vegetable ontological preservation through his experience with the *yakruna*.

Bibliography

Bernales, Albites, "Indigenous Narratives of Creation and Origin in *Embrace of the Serpent*, by Ciro Guerra," *Languages and Epistemologies* 58, no. 1 (2020): 200–213.

Broner, Martina. "Rethinking Format in the Amazon: Ecology and *El abrazo de la serpiente*." *Journal of Cinema and Media Studies* 61, no. 1 (2021): 7–26.

D'Argenio, Maria Chiara. "Decolonial Encounters in Ciro Guerra's El Abrazo de La Serpiente: Indigeneity, Coevalness and Intercultural Dialogue." *Postcolonial Studies* 21, no. 2 (2018):131–53.

Gleghorn, Charlotte. "Filmic Disciples and Indigenous Knowledges: The Pedagogical Imperative in *El abrazo de la serpiente*," *Diálogo* 23, no. 1 (2020): 32–44.

González de Canales Carcereny, Júlia. "Películas en lenguas indígenas producidas en países hispanohablantes: *Ixcanul* y *El abrazo de la serpiente*." *Diálogo* 23, no. 1 (2020): 7–19.

El abrazo de la serpiente, dir. Ciro Guerra. DVD. Colombia: Caracol Televisión S.A.C.V, 2015.

Fueres, Magdalena, Rodrigo Flores, and Rosita Ramos. "Saving Our Seeds: An Indigenous Perspective from Cotacachi, Ecuador." In *Seeds of Resistance, Seeds of Hope: Place and Agency in the Conservation of Biodiviersity*, Eds. Virgina D. Nazarea, Robert E. Rhoades, and Jenna Andrews-Swann, 107–114. Tucson: University of Arizona Press. 2013.

Kohn, Eduardo. *How Forests Think: Toward an Anthropology Beyond the Human*. London: University of California P, 2015.

Morton, Timothy. *Being Ecological*. Cambridge: MIT Press, 2019.

Mutis, Ana Maria. "*El abrazo de la serpiente* o la re-escritura del Amazonas dentro de una ética ecológica y poscolonial," Hispanic Research Journal 19, no. 1 (2018): 29–40.

Salazar, Juan Francisco, and Amalia Córdova. "Indigenous Media Cultures in Abya Yala," In *Media Cultures in Latin America: Key Concepts and New Debates*, ed. Anna Cristina Periterra, 128–146. New York: Routledge, 2020.

Smith, Amanda Mignonne. "From the Rubber boom to *Ayawashka* Tourism: Shamanic Initiation Narratives and the Commodification of Amazonia," In *A Contra corriente*, 14, no. 3 (2017): 1–22.

Treadmell, Holly May. "The Capitalocene in *Embrace of the Serpent*: Capitalism's Destruction of People, Culture, and Nature," *Transmotion* 7, no. 2 (2021): 54–76.

Van Valen, Gary. "A Country Vulcanized: A New Economy," *Indigenous Agency in the Amazon: The Mojos in Liberal and Rubber-Boom Bolivia, 1842–1932*. U of Arizona Press, 2013.

Viveiros de Castro, Eduardo. "Exchanging Perspectives: The Transformation of Objects into Subjects in Amerindian Ontologies," *Common Knowledge* 10, no. 1 (2004): 463–484.

Whitfield, Joey. "Communicating beyond the Human: Posthumanism, Neo-Shamanism, and Ciro Guerra's El Abrazo de La Serpiente." In *Latin American Culture and the Limits of the Human*, edited by Lucy Bollington and Paul Merchant, 177–200. Gainesville, FL: University Press of Florida, 2020.

CHAPTER 7

Enchanting Encounters with Plants in Reshma Aquil and Sumana Roy's Poems

Nesrin Eruysal

In his *Early Writings*, Nietzsche states that consciousness originates from the sensation of causality, whereas memory precedes consciousness. This is exemplified in the mimosa plant, which possesses memory but lacks consciousness, devoid of any imagery. However, memory must be inherent to sensation, a primordial quality of all things, including reflex movements: "Consciousness commences with the sensation of causality, i.e., memory is older than consciousness. For example, in the mimosa we find memory, but no consciousness. Memory of course involves no image in the plant."[1] The sanctity of natural laws affirms that sensation and memory are intrinsic to the essence of all things. Nietzsche also praises the resilience of plants: "What can we find in the agonising labour of all those millions other than the drive to continue to vegetate at all costs: and who does not see the same all-powerful drive in stunted plants stretching their roots into soilless rock?"[2]

Nietzsche's writings on plants resonate well with the plant hieroglyph that Jagadish Chandra Bose, a distinguished Bengali physicist who moved from physics to plant physiology, introduced in his book *Plant Autographs and Their Revelations*. Bose put forward the idea that plants possess memories, respond to touch, and communicate through a hieroglyphic language that must be deciphered and transcribed. The revival of memory is the result of a stimulus thrown on the surface to rouse the dormant images and elicit a fleeting flash of every memory image quiescent in the organism. Bose writes: "These, our mute companions, silently growing beside our door, have now told us the tale of their life-tremulousness and their death-spasm in a script that we can read. May it not be said that their story has a pathos of its own beyond any that we

1 Raymond Guess and Alexander Nehamas, eds. *Nietzsche: Writings from the Early Notebooks* (Cambridge: Cambridge University Press, 2009), 139.
2 Raymond Guess and Alexander Nehamas, eds., *Nietzsche: Writings from the Early Notebooks* (Cambridge: Cambridge University Press, 2009), 68.

 | DOI:10.1163/9789004721777_009

have conceived?"[3] This concept reverberates the Indian belief that everything in the world is interconnected, and life is an endless stream where living and non-living existents merge. The task of interpreting hieroglyphs, transforming the memories of plants into images, and reminding us that all beings, animate and inanimate, are one, falls to both poets and philosophers. Today, poets must undertake the role of creating a new system of values and eco-narratives that reconcile humanity's relationship with the natural world.

In his Introduction to *Plants in Contemporary Poetry Ecocriticism and the Botanical Imagination*, John C. Ryan states that the world is currently experiencing a dramatic moment of reconsideration regarding plants. Drawing on botanical imagination, contemporary poets create a unique imaginative space in which the plant manifests itself and contributes to the poetic process. Ryan probes a series of questions, such as how plants can be imagined and how plants might imagine us, which can be explored through poetry and other artistic approaches. As the imagination extends beyond its normative bounds, poetry becomes a material exploration of the potential of both human and botanical subjects. This destabilizes the ontological distinctions between them, and the imaginative faculty serves as an intermediary zone at the threshold of rational thought and objective knowledge, where plant materiality governs the form and expression of imagination. Ryan labels this new trend of criticism as phytocriticism and gives a brief description of its main tenets:

> The particular mode of botanico-criticism enunciated and put into practice in this study—one that draws from neurobiological standpoints that confer greater agencies and capacities to plants—will be labelled phytocriticism. In acknowledging the plant as a cognitive entity that is able to behave, decide, feel, learn, and remember, phytocriticism assesses the extent to which vegetal dynamism figures into the shape of cultural productions, including environmental texts in which botanical life supplies an active presence.[4]

Through the exploration of the intricate interplay between humanity and plants, poets are bestowed with a diversity of vistas for imaginative vision, transforming the botanical realm into a bountiful repository of images. Here,

3 Jagadish Chandra Bose, *Plant Autographs and Their Revelations* (New York: The Macmillan Company, 1927), 69–70.

4 John C. Ryan. ed., *Plants in Contemporary Poetry: Ecocriticism and the Botanical Imagination* (London, Routledge, 2018), 11.

the botanical imagination assumes the form of an organic process of emergence, a stimulating power that infuses the poetic oeuvre with a vital essence.

Reshma Aquil and Sumana Roy, both distinguished Indian poets, have constructed a poetic universe wherein their observations of human life are intricately interwoven with the natural world. In Aquil's poems, glimpses of her interactions with human and non-human entities, predominantly plants, alongside unarticulated words, and fanciful musings, can be traced. In a similar vein, Roy draws attention to new possibilities that extend from plant to human forms of inscription, which are interwoven into one another. The subtle interplay of geographical, literary, and cultural nuances is skilfully and unobtrusively woven into the fabric of Aquil's poems. On the other hand, the poems of Roy radiate a more intense feeling of West Bengal firmly entrenched in her consciousness.

The first part of this chapter explores the dissolution of the boundaries between humans and plants in Reshma Aquil's poems and includes Emanuele Coccia's arguments that it is possible to create a world if sessile beings can immerse themselves in the world through photosynthesis, which brings all living creatures together. The enigmatic interplay between plants and poetry is also investigated in the light of Gaston Bachelard's argument that childhood is vegetal. Drawing from Vanessa Lemm's article, "What we can learn from Plants about the Creation of Values", I shall further examine how Aquil's poetry teaches us about the importance of values.

In his article entitled "Reshma Aquil of Daryabad," the poet and literary scholar Arvind Mehrotra describes Reshma Aquil as a mysterious individual whose voluntary detachment from the literary community was not merely influenced by her physical location but was rather an inherent aspect of her personality. The topography of Aquil's childhood was dominated by Mahmud Manzil, an extensive single-story house owned by Aquil's maternal grandfather, Agha Ali Khan, a man of high standing and notable pedigree, who held the honourable position of magistrate in the realm of zamindars. The fantastical nature of Manzil's abode was amplified by the presence of its garden and animals, stirring the seeds of poetic inspiration.[5]

Aquil has three poetry collections, *The Unblending, Shadows of Fire,* and *Sleeping Wind.* In *The Unblending,* she explores the struggles of women who rebel against societal expectations and refuse to internalize the limitations imposed upon them. Throughout *The Unblending,* Aquil grapples with issues of

5 Arvind Krishna Mehrotra. "Reshma Aquil of Daryabad." Literary Activism A Series of Presentations and Interventions. Nov 16, 2021. https://www.literaryactivism.com/reshma-aquil-of-daryabad/.

identity, belonging, and acculturation. The collection delves into a plethora of themes, including the complexities of colonial history, the quest for individual and communal enlightenment, as well as the repudiation of ingrained customs. A dialectic between the urge for control and the yearning for liberty ensues, and the poems intimate that subjugation and tyranny arise when individuals seek refuge in controlling their fellow human beings. The presence of plants in the poems is relatively scarce. Only in three poems in this collection, do the botanical images appear. In "Orchids," the delicate flowers, challenge the natural order by piercing through the unyielding rock: "Breaking through rock/ Fragile flower, shoot/ Fear-touched orchids bloom/ Defying an order of rule."[6] The poem lays bare the harsh truths of the established system, whose unyielding dominance can constrain creativity. The orchids, however, serve as a symbol of life's indomitable spirit or the will to power.

The banyan tree in "Banyan" embodies the interconnectedness of life across generations. Its dispersed seeds reveal the uncertainty of existence, but human intervention is necessary to sculpt and confer meaning onto the future. Contrary to the banyan's positive connotations, Aquil's description reveals the inescapable sense of fate and the vicissitudes of life: "But we are seeds/ Scattered far/ Homeless, jobless, undefined."[7] In *The Mind of Plants Narratives of Vegetal Intelligence*, John C. Ryan attempts to delve into the ambivalent mind of the banyan where tree deities dwell. Ryan focuses on the tree's mind: "More than a botanical figment of the imagination, the banyan mind is a breathing phenomenon integral to the Buddha mind—that which finds release from bodily karma and attains awakening."[8] While the tree may elude the cosmic principle of cause and effect, the human species persists in a state of suspectibility: "Without the banyan's/ Ancient guile."[9]

In his book entitled *The Life of Plants: A Metaphysics of Mixture*, Emanuele Coccia elucidates upon the encounters of sessile beings that design their surroundings by passively absorbing the elements of the world around them, consequently constructing an internal realm that seamlessly merges with the external environment. Coccia contends that the process of meeting others is exceedingly intricate for sessile beings including plants and individuals living in solitude, as they lack the capacity to move. He writes: "*To perceive the world* in depth means to be touched and penetrated by it to the point of being

6 Reshma Aquil, *The Unblending* (Writers Workshop, 2003), 18.

7 Aquil, *The Unblending*, 24.

8 John C. Ryan, Patrícia Vieira, and Monica Gagliano, eds. *The Mind of Plants Narratives of Vegetal Intelligence* (Synergetic Press, 2021), 23.

9 Reshma Aquil, *The Unblending* (Writers Workshop, 2003), 24.

changed and modified by it. For a sessile being, knowing the world coincides with a variation of its own form—a metamorphosis provoked by the outside."[10] Yet, plants epitomize the radical expression of existence by possessing the ability to transform the world through photosynthesis. Aquil's poetry exemplifies this immersion that reveals the most radical form of existence. She captures moments of freedom when she distances herself from her personal narrative and delves into the vast realm of literature as a poet and literature professor. Her poems reflect these moments of freedom and encourage readers to transcend their personal histories and explore the depths of their own grounds of being. The solitude that leaves an indelible mark on the poet's soul imbues it with a sensibility that fuels her poetic fantasies and her ability to breathe with plants.

When we study *Shadows of Fire* and *Sleeping Wind*, it becomes evident that in solitude, Aquil has created a world where poetry is the dominant art, and the growth of plants is the means by which time is measured. While the world outside rages with the strife of conflicting religious and cultural practices, her poetry resonates with an intense energy, limiting the influence of the external world to the connections forged with non-human beings. Plants in both *Shadows of Fire* and *Sleeping Wind* embody a direct connection to the world and engage in endless cosmic contemplation. Plants and their interaction with life and festivals such as Moharram processions in the Daryabad region of Allahabad are recurrent themes in these poems that use imagery of plants and their interaction with life to explore themes of transformation, growth, and renewal, highlighting the interconnectedness of all living beings. The poems in *Shadows of Fire* convey the importance of our subjective experiences of the natural world and how they can shape our perceptions and moods.

The poem titled "Breathless" reflects the intermingling of the internal and external domains. The melody of birds, the fragrance of honeysuckles and the hibiscus bush pervade the interior space through a window left ajar: "One tweeting like a pulse from between/The sun-flaming yellow honeysuckle flowers, trembling/All over the railing, you realize/There's no beyond breathless than life."[11] The presence of a cat evokes childhood memories.

"The Encounter" emphasizes the power of natural elements such as wind and rain to disrupt and transform human experience. The palm and gooseberry trees provide shelter, but the sudden arrival of wind and rain highlights the unpredictability of the natural world: "But as I bend to close the

10 Emanuele Coccia, *The Life of Plants: A Metaphysics of Mixture* (Cambridge, UK: Polity Press, 2018), 100.

11 Reshma Aquil, *Shadows of Fire* (Writers Workshop, 2003), 16.

windowpane/Sudden swift winds bring in the rain/Bouncing trees/Frosting everything."[12] The poet's struggle to give shape and meaning to this encounter suggests the limitations of human understanding when faced with the mysteries of nature.

In "Mood Swing," the rainforest is described as a place of light conversation, where the pathway glistens in the dark. The natural world is portrayed as a place full of liberating possibilities, contrasting with the man sweeping the leaves outside. This contrast highlights the importance of our idiosyncratic encounters with nature.

"Changing Hands" speaks to the idea of impermanence in nature. The description of fluttering leaves and flowers suggests that everything in nature is constantly changing. Time is presented as a force that "settles" and then moves on, reinforcing the idea of transience: "Curled leaves, flowers/ Like birds, come and go/ Time settles/ Moves on."[13]

The poet describes the beauty of the mountains and valleys, using phrases like "blue haze of innocence," "cloud-coil streaks," and "white genies ricochet upwards" to create a sense of wonder and enchantment in "White-Out". She also contrasts the natural world with graffiti-covered roofs and rented rooms, emphasizing the fleeting nature of human constructions compared to the perennial beauty of nature: "Mist fleets in between pines like doves/White genies ricochet upwards/In a white-out constant disappear/ Slowly the forest greens/Across a blue stinging sky/A cloud-coil streaks."[14]

In the poem entitled "Shadows of Fire," an unseen realm, shrouded in darkness, alludes to the enigmatic aspects of nature that are capable of both evoking awe and dread. The fire in the woodlands serves as a symbol of how experiences can ignite elemental passions deep within the human psyche. The forest rebel is convinced that breaking her bonds with the natural world will allow her to navigate life's ups and downs easily, never confronting her innermost being. She re-enters the world like Buddha, with a predetermined path and free from earthly desires. Her words erase traces like a forest fire, and she puts on a disingenuous smile: "Cutting through the frill of seas/ Denying the sprawl of tastebuds/ With a smile/ More contagious than the sun's."[15]

In *Sleeping Wind*, the resilience displayed by plants and the feeling of entrapment coexist. Aquil employs the portrayal of a lemon maturing and evolving through the passage of time to delve into the cyclicality of existence

12 Aquil, *Shadows of Fire*, 19.
13 Aquil, *Shadows of Fire*, 28.
14 Aquil, *Shadows of Fire*, 41.
15 Aquil, *Shadows of Fire*, 17.

in "Lemon Tree". The precipitation and soil provide the requisite conditions for growth in "the Autumn's golden bough."[16] The presence of squirrels implies the toilsome endeavour of survival. The depiction of rainy winters cracking the flowers stalks creates a sense of fragility and mortality. However, the difficult blooming of the lemon flowers in the spring implies a sense of resilience and rejuvenation: "In spring's surprise hours/ It is now blooming difficult flowers."[17]

The forest depicted in "Pine Forest" is a living being that breathes and sighs: "We enter the forest/Breathless/Hear the forest sigh!"[18] Still, the symptoms of neurosis manifest themselves in the eerie forest: "And a green cone/ Drops softly behind us/ Like a neurosis, light, wordless."[19] Silence, culturally enforced or self-imposed, can disfigure a life. The forest can teach us to breathe again.

Gaston Bachelard refers to the role that plants play in our mental lives, as sources of inspiration and images that shape our understanding of the world. He argues that we are all influenced by "vegetal values," or the characteristics of plants that are essential to their survival, such as growth, resilience, and adaptability. He suggests that we are all fundamentally connected to the plant world. This suggests a deep reverence for the natural world and an appreciation for the ways in which it has shaped our lives in *The Right to Dream*:

> A botany of the imagination, stemming from an attraction for branches, wood, leaves, roots, bark, flowers and grass, has furnished us with a stock of images of astonishing regularity. We are controlled by *vegetal values*. We would all profit by taking a census of this private herbarium in the depths of our unconscious, where the slow, gentle forces of our life find models of continuity and perseverance. The life of root and bud lie at the heart of our being. We are really very ancient plants.[20]

In *The Poetics of Reverie Childhood, Language, and the Cosmos*, Bachelard urges the reader to consider how the isolation of a dreamer might relate to "the solitudes of childhood."[21] Based on his perception, it is not incidental that we frequently recall the ghostly memories of our childhood. Bachelard writes:

---

16 Aquil, *Sleeping Wind* (Singapore: Ethos Books, 2001), 14.

17 Aquil, *Sleeping Wind*, 14.

18 Aquil, *Sleeping Wind*, 28.

19 Aquil, *Sleeping Wind*, 28.

20 Gaston Bachelard, *The Right to Dream* (Dallas: Dallas Institute Publications, Dallas Institute of Humanities and Culture, 1988), 63.

21 Gaston Bachelard, *The Poetics of Reverie Childhood, Language, and the Cosmos* (Boston: Beacon Press, 1971), 99.

"[t]he vegetable force of childhood subsists within us throughout our lives. Therein lies the secret of our profound vegetalism."[22]

The poems in *Sleeping Wind* show that when we reflect upon our childhood through the lens of its archetypal qualities, it assumes a magnanimous power that transcends our hazy recollections. As the past intermingles with the present, newly fashioned memories infiltrate the tapestry of life that now spans various dwellings. The wind, mountains, and rivers also make their presence felt, engendering a symphony of vitality.

The poem entitled "Star-Struck" brings together terrestrial imagination and aerial imagination. It strikes a harmonious chord with the botanical world and emphasizes its profound significance in the very fabric of our existence. Aquil writes: "After the night has bloomed its garden of stars/And the morning filtered in through the pine,/We stumble down the stairs/Our arms heavy round the pillow/Past the railing, freshened/By the perfume of grapevine flowers."[23] The poem encapsulates the ephemeral essence of the mixed emotions that children experience when they contemplate the vastness of the skies, where fear and fascination intermingle, and they find themselves in a state of profound wonder and awe at the celestial beauty of the stars and the universe. This sensation fades with the emergence of the morning chill. In this cold air and stillness, we witness the formation of a substantial synthesis of air, coldness, and silence that is integral to our being.

The presence of the stars and the wind direction indicate a sense of awareness of the natural world, while the memories of childhood and the racing thoughts reveal the emotional state of the speaker in "Fish-Trail": "A full star and I watch wind directions./The soft gulmohar leaves fan eastward./A sharp scent of jasmine/I slip into childhood."[24] Olfactory memories evoked by gulmohar leaves and jasmines give a sense of motion and direction, thus connecting the natural elements to the movement of time.

In "Shelf Life," the last poem of *Sleeping Wind*, the poet uses the image of insects gnawing on wood to reflect on the ephemerality of human creations. The lines "Nothing stops them/Shredding nations/Puncturing skies"[25] suggest that even the most enduring human structures are vulnerable to the forces of decay and dissolution. Bachelard writes: "In its fiber, the wood always retains memories of its vertical vigor and without considerable skill, one cannot go

22 Gaston Bachelard, *The Poetics of Reverie Childhood, Language, and the Cosmos* (Boston: Beacon Press, 1971), 134.

23 Aquil, *Sleeping Wind*, 9.

24 Aquil, *Sleeping Wind*, 10.

25 Aquil, *Sleeping Wind*, 49.

against the grain of the wood. Thus, for some psychic systems, wood is a kind of fifth element-a fifth matter. It is fairly common, for example, to find wood ranked among the basic elements in Oriental philosophies."[26] The image of swallows nesting in the wood suggests that there is still potential for life and growth.

Vanessa Lemm examines Nietzsche's references to botanical life through the lens of a post-anthropocentric perspective in her article "What we can learn from Plants about the Creation of Values." The post-humanist perspective of the philosopher discerns the human as an integral constituent of the all-encompassing totality of life. In this holistic worldview, the co-existence of animal and plant life is not structured hierarchically, but rather, it is intermingled in a complex web of relationships. Lemm writes: "What plants, animals, and humans share is the freedom to creatively form and transform their form of life."[27] According to Lemm, plants, similar to animals and humans, are beings that generate value, and at the core of Nietzsche's perception of existence lies the very ability to produce values. Nevertheless, Nietzsche's admiration for plants and human plants does not stem from their susceptibility to their surroundings; rather, it is grounded in their ability to thrive and flourish under arduous and unwelcoming circumstances. The plant, rather than being submissive to its surroundings, possesses the aptitude to reformulate its living conditions.

Reshma Aquil's verses are teeming with an implicit struggle between the natural world and the societal frameworks engineered to obstruct individual emancipation. During her childhood, her existence was entwined with the non-human milieu. In later life, as she endeavoured to forge a path of her own, she reignited the bond she had once established with nature. She lived in Daryabad, a confluence of myriad cultures and creeds, as evidenced by the Moharram processions and Christ's blood in the chapel that appear in the poems. The undercurrents of societal turbulence and historical gravity brood in the backdrop, while a yearning for liberty infiltrates the poems. The presence of birds, as exemplified in the poet's words, serves as a poignant reminder of the possibility of flight: "Green in courage I flew again,/Sowing fresh fields, unheeding whistling winds."[28] We can best perceive Aquil's undertaking to generate a value system through the lens of Nietzsche's notions of

26 Gaston Bachelard, *Air and Dreams: An Essay on the Imagination of Movement* (Dallas: Dallas Institute Publications, 2011), 205–206.

27 Vanessa Lemm, "What we can learn from Plants about the Creation of Values," *Nietzsche-Studien* 44, no. 1., (2015): 84.

28 Reshma Aquil, *Sleeping Wind* (Singapore: Ethos Books, 2001), 1.

the will to power latent in plants. Humans can construct values predicated on impressions gleaned from the natural world. Aquil also endeavours to shed light upon the enigmatic interplay between plants and poetry, where they engage in a mutualistic relationship for the purpose of recollecting past experiences. Indian philosophy holds that plants are governed by the principle of karma, where one's deeds in a previous life may result in rebirth as a plant in the next. Therefore, while plants are subject to the law of karma, they do not themselves generate karma, but rather are affected by it.[29] Considering this perspective, we can surmise that the confluence of plant literature and poetry demands that poets undertake the responsibility of transforming the recollections of plants into poetic expressions. Aquil allows plants to speak through the medium of verse.

The second part of this chapter focuses on Sumana Roy's books: *How I Became a Tree* and *V.I.P. Very Important Plant*. Drawing on the rich tapestry of Indian narratives and myths, Sumana Roy, an academic and literary scholar who seeks to decode the languages of plants, explains why she wants to turn into a tree in her book entitled *How I Became a Tree*. Roy wants to have self-confidence and avoid the tactics required for success through the transformation into a tree. She introduces the concept of 'sahrydaya' and discovers that she desires to become a tree one can find in the upwardly oriented tree drawings of Nandalal Bose, whom she describes as a soul mate. The artist's identification with the tree and its effort "to defy gravity"[30] and branch out into the sky is further enhanced by the light. As the narrative advances, Roy turns her attention to the topic of shadows. She argues that, whether they are cast by trees or humans, shadows function as great equalizers that transcend the boundaries of race, class, and religion. Roy also delves into the practices of Rabindranath Tagore and wonders whether Tagore himself cherished a yearning for becoming a tree or flower. The writer explores the karmic states of trees and whether being a tree involves the glorification of sins.

Roy devotes a chapter to the achievements of Jagadish Chandra Bose and shares her impressions of his house in Darjeeling. She refers to a letter written by Bose to Rabindranath Tagore, where Bose reflected on his life in Darjeeling, characterized by a serene environment where not a single human voice could be heard. Bose's writings, published in Bengali and entitled "Awbyakto", meaning "the unsaid", express his deep sense of resentment. In his view, plants too should have their own language to tell their stories. He calls this language or

29 Ellison Banks Findly, *Plant Lives: Borderline Beings in Indian Traditions* (Delhi: Motilal Banarsidass, 2008), 159.

30 Sumana Roy, *How I Became a Tree* (New Haven & London: Yale University Press, 2017), 37.

plant-writing, "Torulipi". Bose's book *Plant Autographs and Their Revelations* has challenged traditional views on the capabilities of plants, forging novel paths of inquiry within the realm of plant biology. The book delves into the intricate electrical and mechanical properties of plants, elucidating how these properties can be precisely measured and recorded. Roy writes:

> Bose's frustration and disappointment is with the 'silence', the lack of reciprocal communication, his need for a dialogue with plants. So conditioned are we to a back-and-forth system of speaking that we do not realize that there could be alternative models of communication as well. The essay 'Nirbak Jeebon', The Silent Life, begins with this tautology: The life around us is terribly silent. And then the old desperation—we have to get plants to tell us their own life stories. These histories must not be mediated by man, they must come to us directly. Bose's insistence on the 'direct speech' of plants, without man's 'prejudiced intentionality', when it prefaces his comparison between plants and infants, wouldn't strike us as crazy if we think of the many times philosophers, linguists, and brain scientists have wanted to get inside the minds of infants and toddlers to become aware of their thoughts. That is also the reason why all 'histories of childhood', like the histories of plants and trees, are incomplete.[31]

Seeking to escape the clamour of human society and the modern consumer-driven lifestyle, Roy decides to journey to the forest in search of her true desires. The forest is akin to a venerated Hermes, the ambivalent messenger of Gods. Roy writes: "The forest is a Hermes-like figure, moving between states, a figure of transitions and translations. Planting trees is the only way to kill death, to achieve immortality, a fact Jugalprasad recognizes."[32]

In the later chapters, Roy's exploration of Indian beliefs uncovers the significance of the tree as a symbol of liberation. Taking refuge under its branches grants access to a realm of sacredness, where one can escape the allure of greed and the mundane. Buddha exemplifies this, achieving enlightenment under the Bodhi tree, rooted in the divine Brahman. The focus slowly shifts towards an introspection of the karmic cycles and the ultimate annihilation of trees. Ellison Banks Findly probes the Jain doctrine which postulates that an individual may undergo reincarnation as a tree and possess the faculty to understand the agony that is attendant upon physical harm caused by touch.

31 Roy, *How I Became a Tree*, 135.
32 Roy, *How I Became a Tree*, 175.

This ontological position indicates that plants too are susceptible to karmic outcomes such as elation, melancholy, anxiety, and wrath depending on the karmic debt. These experiences "are the resultant effects of the ever-going karmic accumulation associated with the embodied soul."[33] As Roy contemplates the possibility of such transformation as an escape from human cruelty, she acknowledges the inevitability of suffering and fervently wishes to transform into a tree that is immune to the anguish of existence: "I'd choose to be an Ashoka, the a-shoka, the sorrowless tree, I decide."[34] Findly writes:

> Another story in Mahabharata describes the Asoka tree as a tree that can free people from their sorrows … The Asoka tree, whose name means "without sorrow" and "is believed to be a remover of sorrow" is one of those trees of the subcontinent who has been "treated like human beings, endowed with a soul, a heart that weeps with grief and laughs with joy." It is sacred to Hindus and Jains. Jain thinkers attribute life and karma to plants.[35]

Roy has laboured to reconcile herself with the arboreal chronology, yet she remains ambivalent as to the efficacy of her metamorphosis into a tree. However, when a bird alights upon her shoulder, any hesitations that once plagued her vanish, and she finds herself resolute in her decision to take on the appearance and soul of a tree. Finally, she evokes the image of a bird to underscore the deep interdependence of birds and trees. Their symbiotic relationship ignites her desire to transform into a tree:

> Like trees I wanted no more than I needed; I imagined my relationship with light as akin to that between it and trees; I tried to live to tree time, rejecting speed and excess. And yet I did not feel completely like a tree.
>
> Not until a bird came and sat on my shoulder around sunset one day. I did not move. I do not know about the bird but I was certain that in the thinning margins of that forest in Baikunthapur I was, at last, ready to be a tree.[36]

In his seminal work "Plant-Thinking: A Philosophy of Vegetal Life," Michael Marder posits the query of how to encounter flora, which is often relegated to the periphery of philosophical discourses. The presence of plants engenders

33 Findly, *Plant Lives: Borderline Beings in Indian Traditions*, 159.
34 Roy, *How I Became a Tree*, 228.
35 Findly, *Plant Lives: Borderline Beings in Indian Tradition*, 156–157.
36 Roy, *How I Became a Tree*, 220–221.

a curious sense of unfamiliarity. Marder writes: "Curiously enough, the absolute familiarity of plants coincides with their sheer strangeness, the incapacity of humans to recognize elements of ourselves in the form of vegetal being, and, hence, the uncanny—strangely familiar—nature of our relation to them."[37] Marder's search for answers to his inquiries resonates well with the poems included within Sumana Roy's poetry collection: "V.I.P. Very Important Plant," where the boundary between humanity and plants becomes increasingly blurred. The poems in this collection transcend the limitations of anthropocentric thought by tracing the interconnections that exist within all living things and seeking to discover the desire for growth and loyalty to earth among plants. The bond between plants and humans is not limited to vitality but rather extends into the realm of mortality.

The death of a human being differs markedly from that of a plant. No ceremonies or prayers offer solace to the departed plant. The plant, like humans and animals, lived a life in which it struggled to escape from samsara, that is, the cycle of death and rebirth. In what body will the reincarnation bring her back to life when she is brought back to life by karma? Based on these questions, considering Bose's observation that life continues in some way after death, perhaps it can be said that there is no need for consolation just because there is no death. Roy writes in "How to Console a Dying Plant": "What use are the Scriptures or the drumbeat of last wishes to a plant? / Or a hand and its show of genres-patting, holding, and then leaving?"[38]

In the poem entitled "I Want to Be a Tree," the poet conveys her intention to lead a nonconformist life that challenges the strictures of societal norms. Though sheltered from the oppressive grasp of the city, the suffocating blanket of summer's heat weighs heavy upon the leaves. The winds, a force that resists being confined to the written page, carry the flame-filled book of summer's wrath. Love, rather than religion, becomes the narrator's aspiration, as it is an emotion that endures even beyond death and continues to be expressed to those who have departed: "Love needn't be reciprocal-/ How else can we love the dead?/ The earth loses ownership of dead trees."[39] The loss of chlorophyll, the pigment responsible for the hues of plants and their ability to perform photosynthesis, may symbolize the loss of one's vibrancy and vitality within a relationship.

---

37 Michael Marder, *Plant Thinking: A Philosophy of Vegetal Life*. (New York: Columbia University Press, 2013), 4.

38 Roy, *V.I.P. Very Important Plant*, 19.

39 Roy, *V.I.P. Very Important Plant*, 24.

"Tree and Air" displays the interplay between storytelling and trees. The parched and withered leaves that fall from trees cannot tell a tale, for everything in existence is at the mercy of the all-encompassing air. When the narratives that chronicle life are shared, they dissolve into the atmosphere, perhaps dissolving into the collective unconscious. The storyteller metamorphoses into a tree; storytelling, therefore, serves as a means of becoming a tree: "When you tell a story, you become air./ As reckless as thyme, as nagging as health./ When you finish a story, you become a tree."[40]

When the poet yearns for trees that are translucent, she seeks refuge in the trees' multifaceted existence. In "Eating Light," the shadows never disappear despite growing up and gaining a grasp on the world's objective reality. On the contrary, they grow alongside the individual, becoming so pervasive that one may aspire to transform into a tree to flee from them. Growing up, one sees the world as it is, in plain reality, but this does not mean that the shadow has disappeared: "This hunger for light, this for becoming tree./ When a sapling grows, so does its shadow."[41]

The lines in "Fasting Without Light": "The tree roots end their daily pilgrimage,/ the chorus of grass rebels against fonts of tidiness,/ and the tree rests, at last, with its dead shadow,"[42] encapsulate the uncanny interplay between the two opposing forces of light and darkness. The slow disappearance of shadows from our consciousness is similar to the feeling of estrangement. We are often convinced of the permanence of sunlight or consciousness when it first appears, but this is merely a misleading delusion. As the evening descends, a melancholic air descends with it, as one yearns to settle in. The plants become playthings, and the fields remain dependent on the sun's sustenance. The distant moss appears offended, for the sun's radiance can't penetrate its secluded corners. Although voices abound, we remain deaf to their utterances.

"The Afterlife of Trees and Their Lovers" tells of Roy's visit to Bose's house in Darjeeling, where the poet goes to learn the language of plants. Ghosts of Bengali trees, or ghostly beings accompany this journey: "It is difficult to imagine a history of trees/ without man in it./ Man as tree, Tree as tale."[43] The poet wonders if the trees could remember Bose. It is noted that botanical science deals with the personality of plants. When considering Bose's work on understanding the language of plants and his ideas about it, it's important not

40 Roy, *V.I.P. Very Important Plant*, 29.

41 Roy, *V.I.P. Very Important Plant*, 39.

42 Roy, *V.I.P. Very Important Plant*, 42.

43 Roy, *V.I.P. Very Important Plant*, 43.

to overlook his spiritual beliefs and the changes in his spiritual worldview. The intertwining of trees and spirituality is evident in Bose's life story.

In the second part of the poem, Shakti Chattopadhyay's house in Baharu is visited. Chattopadhyay was an Indian poet and writer who wrote in Bengali known for his realistic depictions of rural life. He was a green poet, and many of his poems raised the issue of nature in crisis. Through his poems, he urged people to protect Mother Nature and plant trees. Every tree tells a story. The last part of the poem is about Bodhi Tree. Individuals who elect to dwell in seclusion amongst the verdant foliage of the forest—be it the forest rebels, or the legendary personas of Rama and Siddhartha who sought inner tranquillity—find solace in its serenity. Gautama Buddha, too, discovered the futility of the corporeal form and the limitations of language through his contemplation under the tree's shade: "Only I know that the tree is Buddha./And that the Buddha was a tree."[44]

As one enters the third stage of life, one should retreat to the forest. It is an arena that cultivates introspection and communion with nature. The completion of home-building tasks signifies the arrival of the stage of human contemplation. To traverse the woods is to enter an ethereal memoir, where trees exist as neutral entities, beyond the confines of human assessments of success or failure. Roy writes: "To be inside a forest is to be part of an invisible autobiography."[45]

"Root Vegetables" and "Roots" emphasize that death is compulsory, and our lives get entangled with the leaves inside the earth. The poet, in her pursuit of conveying a more comprehensive understanding of botanical life, exhibits a remarkable propensity not only for delineating her impressions of plants and soil but also for delving into the realm of death. "Tagore's Trellis" incites contemplation of the appellations of certain plants, inspired by the tradition of endowing them with new names. Naming holds therapeutic properties: "You renamed plants like a doctor changing a drug. / Names, their feet-dragging newness, would be cure."[46] In Part IV of *How I Became a Tree* which begins with a quote from Rabindranath Tagore, Roy also expounds upon the rationale behind Tagore's proclivity for endowing his leafy companions with new names, thereby executing a ceremonial initiation.

In "Algae" Roy employs words that evoke negativity—such as death, ignorance, foreignness, panic, pain, and vulnerability to depict the algae in the British botanical artist, collector, and photographer Anne Atkins's photograms:

---

44 Roy, *V.I.P. Very Important Plant*, 46.

45 Roy, *V.I.P. Very Important Plant*, 49.

46 Roy, *V.I.P. Very Important Plant*, 63.

"as smooth as death, as calm as the ignorant".[47] There is an aura of mortality pervading the atmosphere. The spectral essence of botanical life begets an eerie disquietude, resonant in Atkins's photograms. This poem provides a foundation upon which to conduct an analysis of the relationship between plant writing and photography. Patrícia Vieira's essay entitled "Phytographia Literature as Plant Writing" explores the correspondences between phytographia and photography. She states that a comprehensive comprehension of the concept of phytographia can be achieved through a thoughtful correlation with the art of photography. In essence, photography can be likened to the act of engraving light, where the physical nature of objects commingles with luminosity, ultimately resulting in a representation of reality as interpreted by the photographer's creative vision. Similarly, plants already engage in proto-photography, as evidenced by their inscription within the environment through the process of photosynthesis. Vieira writes: "I see literature, the realm of the imagination, as a mediator in the aesthetic encounter with plants, knowing full well that the medium always colours the message and that the mediators themselves are all but evanescent."[48]

In "Weeping Willow," the reader is enlightened to all the wonders that can be learned from a tree—from the mere dust and wind to the roots that endure through times of temporary hunger and the plaintive cries of the stars. It is the account of a tree remembered through the veil of tears that ought to be hidden away like the roots that grip the soil: "You turn to trees to look for the same-/the dust of elsewhere in the freckled wind,/ the anger of temporary famines in their roots,/the complaint of stars in the light travelling from the sun."[49]

In *Air and Dreams Nietzsche and the Ascensional Psyche*, Gaston Bachelard asks how many legends have shown us the tree that weeps and delivers a penetrating reflection: "The suffering tree is the epitome of universal pain."[50] However, one may not disregard the probability of the tree's suffering is a manifestation of its karmic debt, a concept that transcends the anthropocentric construct of pain. The suffering tree, then, is an emblematic representation of universal anguish, a potent reminder that every life form, no matter how different, is bound by the inextricable laws of the universe.

47 Roy, *V.I.P. Very Important Plant*, 66.

48 Patrícia Vieira, "Phytographia Literature as Plant Writing," in *The Language of Plants Science, Philosophy, Literature*, eds. Monica Gagliano, John C. Ryan, and Patrícia Vieira, 215–233 (Minneapolis: Minnesota University Press, 2017), 225.

49 Roy, *V.I.P. Very Important Plant*, 72.

50 Gaston Bachelard, *Air and Dreams: An Essay on the Imagination of Movement* (Dallas: Dallas Institute Publications, 2011), 217.

"Do Trees Hurt?" describes the plight and ultimate salvation of Lazarus. The seeds return every year, like Lazarus, who was resurrected after his death. The stratified textures of trees bear witness to the years, but not to the anguish and sorrows that befall them. Humankind's memory records these tribulations and secrets, yet never forgets. The layered textures of trees record years, but do not remember fears and hurts. Roy wants to change the course of karmic laws in this poem:

> that is why you want to be a tree-
> to be free of injuries, of servility,
> of beds that hold the foetus of your fears.
> its cambium record years, time as vermicelli,
> but not hurt's ageism, not grief that clogs truth and rivers.[51]

"Potato" presents fragments from a childhood memory, resonating with Bachelard's assertion that the youthful phase is akin to the vegetative state. The harvest of potatoes in her grandfather's village and the attention given to the potatoes, where the fecundity of the soil is concentrated, represent the plants that hold a dominant role in the recollections of childhood. Potatoes where the talent of the soil is concentrated represent the plants that play a dominant role in childhood memories: "Until the soil explodes and the potatoes emerge/ like grooms./ The elasticity of our joy bursts./ We fall to the ground, fattened by laughter."[52]

The Asvattha tree holds a revered status among Hindus, as it has been praised in various texts pertaining to both Hinduism and Buddhism. In fact, the Bodhi Tree, which played a vital role in the enlightenment of Gautama Buddha, belongs to the same species. As the daylight fades, the leaves of this tree relinquish their hold, thus drawing attention to the plant-specific emotional transformations that Bose highlighted. The descent of nightfall, the passing of clouds, and other natural phenomena all cause the plant to shudder and experience an unsettling restlessness. In "Asvattha," Roy explores the origins of the tree's name and thinks that the leaves of the tree, meaning the tree under which the horses stand, are rehearsing for a storm, and suggests that very moment is preparing us for the ultimate demise. "The leaves are moving, the light is dying,/ they are surrendering to rambling darkness,/ to its intimacy

51 Roy, *V.I.P. Very Important Plant*, 85.
52 Roy, *V.I.P. Very Important Plant*, 88.

which preserves constancy."[53] Although the branches move, they continue to stay where they are, like the poet who has nowhere to go.

The poet writes about plants such as peach, apple, garlic, fern, malatilata, lemon, mint and papaya among many others to make their voices heard and takes on the task of a messenger like Hermes. She oscillates between the plant world and the human world and delves into the world of the plant to hear: "a hardened memory of your journey inside soil."[54] From time to time, she makes surprising analogies like Hermes, deliberately or unknowingly, interfering with the content of the messages. "Turmeric," the last poem in the book changes form with each rebirth and preserves its memories under the ground. When examining the roots of the term, one discovers a depiction of *Terra marita* or the fertile ground for imagination. The same plant has fifty-two names in Sanskrit. These names carry traces of darkness, earthy memories, and blindness. Roy ends the poem claiming that where the Westerner saw talent and virtue, the Indian saw the illusion of Maya.

Vanessa Lemm portrays Nietzsche's plants as sentient beings possessing a faculty of measurement and historical consciousness. Lemm argues that this does not require a complete rejection of Anthropos and that humans and plants should exist symbiotically within a new paradigm of world perception, allowing them to learn from each other and create artistic expressions and values. Both Aquil and Roy enhance the move away from anthropocentric to plant-based poetry enriched by botanical imagination. However, they are aware that our understanding extends far deeper into the liberating nature of things. This understanding can aid us in transcending our current state and surpassing the limitations that hinder the full development of our individual potential.

Through their investigations into the botanical imagination, Aquil and Roy aim to discover unexplored terrain for poetic innovation and encourage contemplation of the intricate ties that bind humans and plants. Their works emphasize the significance of myths, folk tales, and religious rituals in shaping people's approach toward the entanglement of human and non-human elements. Furthermore, they show that the experiences of childhood and proximity to nature during the formative years of one's life play a pivotal role in establishing environmental sensitivity, thereby augmenting awareness of environmental predicaments.

---

53 Roy, *V.I.P. Very Important Plant*, 93.

54 Roy, *V.I.P. Very Important Plant*, 97.

Bibliography

Aquil, Reshma. *The Unblending*. Writers Workshop, 2003.

Aquil, Reshma. *Shadows of Fire*. Writers Workshop, 2003.

Aquil, Reshma. *Sleeping Wind*. Singapore: Ethos Books, 2001.

Bachelard, Gaston. *Air and Dreams: An Essay on the Imagination of Movement*. Dallas Institute Publications, 2011.

Bachelard, Gaston. *The Right to Dream*. Dallas: Dallas Institute Publications, 1988.

Bachelard, Gaston. *The Poetics of Reverie Childhood, Language, and the Cosmos*. Boston: Beacon Press, 1971.

Bose, Jagadish Chandra. *Plant Autographs and Their Revelations*. New York: The Macmillan Company, 1927.

Coccia, Emanuele. *The Life of Plants: A Metaphysics of Mixture*. Cambridge, UK: Polity Press, 2018.

Findly, Ellison Banks. *Plant Lives: Borderline Beings in Indian Traditions*. Delhi: Motilal Banarsidass, 2008.

Geuss, Raymond, and Alexander Nehamas, eds. *Nietzsche: Writings from the Early Notebooks*. Cambridge: Cambridge University Press, 2009.

Lemm, Vanessa. "What we can learn from Plants about the Creation of Values." *Nietzsche-Studien* 44 no. 1, (2015): 78–87.

Marder, Michael. *Plant Thinking: A Philosophy of Vegetal Life*. New York: Columbia University Press, 2013.

Mehrotra, Arvind Krishna. "Reshma Aquil of Daryabad." Literary Activism A Series of Presentations and Interventions. Nov 16, 2021. https://www.literaryactivism.com/reshma-aquil-of-daryabad/.

Roy, Sumana. *How I Became a Tree*. New Haven & London: Yale University, 2022.

Roy, Sumana. *VIP Very Important Plant*. Swindon: Shearsman Books, Press, 2017.

Ryan, John C., ed. *Plants in Contemporary Poetry: Ecocriticism and the Botanical Imagination*. London, Routledge, 2018.

Ryan, John C., Patrícia Vieira, and Monica Gagliano, eds. *The Mind of Plants Narratives of Vegetal Intelligence*. Synergetic Press, 2021.

Vieira, Patrícia. "Phytographia Literature as Plant Writing" In *The Language of Plants Science, Philosophy, Literature*, eds. Monica Gagliano, John C. Ryan, and Patrícia Vieira, 215–233. Minneapolis: Minnesota University Press, 2017.

CHAPTER 8

Flowering-with the Silence of Plant: Affective Attuning to the Vegetal Event in Tang Poetry

Perkus Chun-kit Leung

1 A Search for a Plant-Voice in Chinese Poetry

Magnolias blooming self-so-ly in tranquility, cassias falling leisurely in quietude, stream flowers flowering affectively in quiescence: in its florescence and senescence, affecting and being affected in the milieu of interrelations, a plant *voices*, in profound eloquence, an affective silence that, when gesturing a detour from human subjectivity into the realm of unspoken agencies, plants a moment of eventfulness in its bloom. In his "What's Planted in the Event?", Michael Marder traces the event to the scenario of an "out-coming" inhered in its Latin etymology *ex-venire*, and expressively defines the concept as a way of "coming-out": "out of the womb, out of the world or out of life, or, again, out of one's isolated self to meet the other."[1] Deep in the silence of the vegetal event is a plant-voice that expects a symphony with myriad beings. It endows us with an alternative plant-thinking that scrutinizes anthropocentric discourses under the promulgation of the Anthropocene which validates the human agency as a sole voice territorializing the planet's geology and ecology. In Chinese poetry, particularly the landscape poetry composed in the Tang dynasty, the plant world is a manifestation of the Buddhist teachings of "purifying the heart and illumining one's true essence 淨心明性" and the Daoist ideal of self-so-ness 自然.[2] More specifically in relation to the rethinking of the Anthropocene with plants, vegetal silence in Chinese poetry becomes a germinative site that echoes Marder's notion of a genuine plant-thinking as adumbrated in *Plant-Thinking: A Philosophy of Vegetal Life*: "let plants maintain their otherness" while "respecting the uniqueness of their existence."[3]

1 Michael Marder, "What's Planted in the Event? On a Secret Life of a Philosophical Concept," in *The Green Thread: Dialogues with the Vegetal World*, ed. Patrícia Vieira, Monica Gagliano and John Ryan (Lanham, Boulder, New York & London: Lexington Book, 2016), 3.

2 Pan, Fuh-jiunn, *Botany in Tang Poetry* (Taipei: Owl Book, 2018), 205.

3 Michael Marder, *Plant-Thinking: A Philosophy of Vegetal Life* (New York: Columbia University Press, 2013), 8.

 | DOI:10.1163/9789004721777_010

It is impossible to overstate the significance of plants to Chinese poetics, be it owing to the preponderance of vegetal imageries, or to the fact that many poets drew inspiration from plants. As documented in Pan Fuh-jiunn's 潘富俊 *Botany in Chinese Literature*《中國文學植物學》, over 50% of the Chinese classical poems make reference to plants.[4] Since the publication of *The Book of Songs*《詩經》, plants have been an essential part of early Chinese literature, functioning often as ideal vehicles for a euphemistic expression of emotions particularly in Chinese poetry. This admiration of plants, as Donald Harper notes in "Flowers in T'ang Poetry," denotes the Tang poets' "strong empathetic attachment to flowers, especially to those flowers that were cultivated in gardens",[5] as a result of their cognizance of the resemblance between plants and real-life vicissitudes.

It is a pity, however, that critics approached plants superficially as mere metaphorical expressions. It is up to later critics, in our case Christopher K. Tong and Wong Kin-yuen, to develop a non-anthropocentric nonhuman Chinese poetics. In his "Nonhuman Poetics," Tong proposes an object-oriented poetics that perceives objects by means of objects 以物觀物, according to which "nonhumans do not merely serve as surrogates for human affairs but rather enjoy an autonomous existence beyond human signification and identification."[6] Similarly, Wong envisages a plant-favored poetics in his "The Treeing-of-tree through Affective Attunement," in which he illuminates on the agency of plants—delineated, namely, by referring to vegetal morphemes in the poetic diction—in co-weaving an "interspecies ethics and aesthetics".[7] Later in "As the Tree Greens," Wong resorts further to the event "to green" developed in the philosophy of Gilles Deleuze, for reinscribing upon the ontologies of plants the regenerative force of incorporeal transformation of births and rebirths, "the living force of life giving all other forms of life an ongoing continuity ... the very drive to strive."[8] The event of reencountering plants is to return the botanical beings to their ontologies, their silence. Interestingly,

---

4 Pan, Fuh-jiunn, *Botany in Chinese Literature* (Taipei: Owl Book, 2011), 16.

5 Donald Harper, "Flowers in T'ang Poetry: Pomegranate, Sea Pomegranate, and Mountain Pomegranate," *Journal of the American Oriental Society* 106, no. 1 (1986): 149.

6 Christopher K. Tong, "Nonhuman Poetics (by Way of Wang Guowei)," in *Ecocriticism and Chinese Literature: Imagined Landscapes and Real Lived Spaces*, ed. Riccardo Moratto, Micoletta Pesaro and Di-kai Chao (London & New York: Routledge, 2022), 44.

7 Wong Kin-yuen, "The Tree-of-tree through Affective Attunement: Biosemiotics and Chinese Ideograms as an Ecosystem," *Ecolinguistica: Revista Brasileira de Ecologia e Linguagem* 6, no. 2 (2020): 45.

8 Wong Kin-yuen, "As the Tree Greens: Deleuze's Form-Event Assemblage and Chinese Ideograms in a Biosemiotic Ecosystem," *Deleuze and Guattari Studies* 15, no. 2 (2021): 304.

albeit Wong has not delved much into the nuance of silence, he is constantly addressing us to be attuned to the profound tranquility of plants by which the plant life is "granted agency of expressing itself."[9]

It is therefore the aim of this chapter to examine the silence of plants in Chinese poetry in search for a lively plant-voice that resists Anthropocenic discourses about anthropocentrism and humanism. To adumbrate the scenario of attuning to the silence of vegetal life, this paper takes Marder's idea of a vegetal event as a conceptual threshold to a genuine perception of the botanical rhythms and agencies through affect. This paper resorts also to the Buddhist idea of emptiness and the Daoist notion of non-action for proffering a non-Western mode of plant-thinking. Through examining the non-anthropocentric portrayal of plants in selected Tang poems, this paper argues that an eventful encounter with plants resides in an attunement to the lacunae and silences of plants, wherein the vegetal silence emerges as an inter-rhythm that orchestrates our imbricated rootedness to this kaleidoscopic world of non-human agencies.

2 Affect, Vegetal Event and the Chinese *Huā*

The idea that a plant expresses silence is not a paradox if we redefine the concept of "voice" to include nonverbal forms of expression. By considering a plant's posture as a form of (bio)semiosis, we can see it as articulating its active engagement with its environment. This broader understanding allows us to recognize the agency of plants in their interactions with the world around them. The notion of a plant-voice is introduced by John Charles Ryan in his *Posthuman Plant* to deliver a robust criticism against the audiocentric logos that privileges a human-centred and monologic notion of voice, according to which "a plant voice must speak if it is to be heard; if it does not speak, it is mute, an utterance of no consequence, with neither register nor agency."[10] Our difficulty in perceiving plant-voice is twofold. First, as Randy Laist notes, the "utterly alien" morphology of plants conceptually renders them "easy and even tempting to deny their status as animate organisms."[11] Second, our misunderstanding of difference as negativity begets the confusion of silence with

9 Wong Kin-yuen, "Tree-of-tree," 41.

10 John Charles Ryan, *Posthuman Plants: Rethinking the Vegetal through Culture, Art, and Poetry* (Illinois: Common Ground Publishing, 2015), 140.

11 Randy Laist, "Introduction," in *Plants and Literature: Essays in Critical Plant Studies*, ed. Randy Laist (Amsterdam & New York: Rodopi, 2013), 12.

muteness. That is, as Marder constantly reiterates throughout his works, any mode of articulation "undetectable on the radars of our conceptualities"[12] is deemed as a form of incompetence and incapability, if not deficiency, despite the fundamental differences among species in terms of physiology.

It is therefore Ryan's call to "think about voice differently,"[13] by reimagining a voice "not of the animalistic kind" but that which "perfectly corresponds to its ontology and habitus, as the plant finds a way of announcing itself."[14] As Ryan argues later in *Plants in Contemporary Poetry*, the spirit of our plant-thinking is to salvage a lively plant—the one "that lives, one in possession of a life, one that expresses and enacts its subjectivity within a life-world" without being "homogenizable as the blunt biological denomination *plant life*" or "negativized in the image of animal life and human life."[15] As nonhuman and non-animal life forms, plant-voice encompasses both an affirmation of the utter otherness of the vegetal life, and an abandonment of "the familiar terrain of human and humanist thought".[16] This recognition of the vegetal agency is vividly described in the last couplet of Zhang Jiu-ling's 張九齡 "Contemplating 2" 《感遇2》: "The grass-woods, having their own hearts,/ Why do they need to be picked by a beauty? 草木有本心，何求美人折。"[17] With the acknowledgement of the ontology of grass-woods, the couplet critiques the literary tradition of representing plants as a metaphor for beautiful females in Chinese culture, which is nonetheless an anthropocentric practice of reinforcing the familiar terrain of human values. As both Ryan and Marder would agree, the vibrant plant-voice dwells not in a plant's confirmation to human logos but its embodied and embedded relationalities to its milieu. Deep in the silence that defies interference, the plant life voices a symphony with the environment in the events of the flowering of flowers, the blooming of trees, the withering of petals—in affect.

The notion of affect is germinative in the context of plant-thinking insomuch as it pertains to "a body's *belonging* to a world of encounters" or "a world's belonging to a body of encounters";[18] in short, the embodied relationality

12 Marder, *Plant-Thinking*, 12.
13 Ryan, *Posthuman Plants*, 142.
14 Ibid., 147.
15 John Charles Ryan, *Plants in Contemporary Poetry: Ecocriticism and the Botanical Imagination* (New York & London: Routledge, 2017), 16.
16 Marder, *Plant-Thinking*, 10.
17 Wong Kin-yuen, "Treeing-of-tree," 54.
18 Gregory J. Seigworth and Melissa Gregg, "An Inventory of Shimmers," in *The Affect Theory Reader*, ed. Melissa Gregg and Gregory J. Seigworth (Durham & London: Duke University Press, 2010), 2.

between a body and its milieu. Affect as a concept can be traced back to works by Baruch de Spinoza (1632–1677). In *The Ethics*, Spinoza defines affect as "affections of the body by which the body's power of acting is increased or diminished, aided or restrained, and at the same time, the ideas of these affections."[19] In words by Brian Massumi, affect names a relational event of reciprocal transformations, itself being a "life-filling, life-forming journey."[20] It is a lived embeddedness of an intermingling *com-posure* of reciprocal sensations and relationalities. As affect, Ryan suggests that the notion of voice can be "extended to linguistic and nonlinguistic acts alike, leading to ethical implications for the rhetor and the receiver."[21] Conceiving voice through the lens of affect, a plant-oriented voice expresses the vitality of a plant through its situatedness and relationality with others; in Marder's words, its flowering-with "the world's fluctuations, seasonal rotations, comings and goings."[22] The notions of vegetal affectivity and embodiment are expressively inscribed on the Chinese ideogram of the character *huā* 花 (flower) (fig. 8.1). Etymologically, the small seal script of the character is a pictorial depiction of a flower stretching its petals and leaves, and its roots spreading deep in the soil, itself being a manifestation of the *zào huà* 造化—creation and transformation—of differential elements.[23] The flowering of plants is a "syncretism of their environmental relations",[24] a continuity with their milieus, and a fruit of their affective resonance with rhythms of other bodies.

The subtlety of the Chinese character *huā* resides in its manifestation of the metamorphosis, or rather, eventfulness, of botanical beings, which is made more explicit in its modern regular form (fig. 8.2). Morphologically, *huā* is a composite of a vegetal radical (艹) and *huà* (化), which, when taken as a single Chinese character, graphically portrays one person standing upright and another upside down, and thus articulates the meanings of subversion, dynamism and transformation. In and of itself, the ideogrammatic design of *huā* weaves a scenario of a vegetal dynamism—more precisely, a vegetal event—that manifests a self-so becoming in sync with the milieu that transcends subjectivity and the resulted dichotomy between the self and the world.

19 Benedict de Spinoza, *A Spinoza Reader: The Ethics and Other Works*, trans. Edwin Curley (Princeton: Princeton University Press, 1994), 154.

20 Brian Massumi, *Politics of Affect* (Cambridge: Polity, 2015), viii.

21 Ryan, *Posthuman Plants*, 146.

22 Michael Marder, *The Philosopher's Plant: An Intellectual Herbarium* (New York: Columbia University Press, 2014), 222.

23 Gu Jian-ping, *A Graphic Compendium of Chinese Characters* (Shanghai: Orient Press Center, 2008), 221.

24 Ryan, *Posthuman Plants*, 152.

FIGURE 8.1 Small seal script 小篆 of *Huā*

FIGURE 8.2 Regular script 楷書 of *Huā*

The notion of *huà* is of particular importance in Daoist philosophy, specifically in the Daoist classic *Zhuangzi*《莊子》. In the chapter "On the Uniformity of All Things,"〈齊物〉the notion of *wùhuà* 物化—"the transformation of things"[25]—is introduced to address the obliteration of subjectivity and the transcendence beyond dualisms between subjects and objects—and hence, "I" and the world—in the event of mutual imbrication among all things as indicated by the title of the chapter.[26] Likewise in the chapter "Let Be and Let Alone,"〈在宥〉the character *huà* is reiterated to designate this Daoist practice of self-obliteration through "[f]ollow[ing] the natural course of events and do[ing] nothing. 汝徒處無為，而物自化"[27] With the implication of a self-so transformation that unfurls an affective attunement to other entities, the *huà* in *huā* affords a vegetal event rooted in its embodied imbricatedness within the milieus crisscrossed by trajectories and rhythms of myriad life forms.

The spirit of an earthed metamorphosis inhered in the regular script of *huā* resonates with the scenario of a situated "out-coming" adumbrated in Marder's conceptualization of the vegetal event. With an emphasis on the suffix *ex-* in the Latin etymology of the word "event" (*ex-venire*), the concept designates the "spatial articulation in excrescence, temporal articulation in expectation, and freedom from rigid determination in exception ... on the basis of plants."[28] Precisely as the germination of its continuity and contiguity with its elemental

25 *Zhuangzi*, trans. Wang Rongpei (Hunan: Hunan People's Publishing House, 1999), 41.

26 *Zhuangzi* is famous for its writing style which often includes the use of parables. When introducing the concept of *wùhuà*, there is a parable of master Zhuang's dream in which he becomes a butterfly who is joyously immersed in its fluttering, forgetting that it is master Zhuang itself. More interestingly, when master Zhuang wakes up, he is uncertain whether it is that master Zhuang is dreaming of the butterfly, or that the butterfly is dreaming of master Zhuang. The notion of *wùhuà* is used to conceptualize such an eventful blurring of the boundary between subjects and objects.

27 Ibid., 167.

28 Marder, "What's Planted in the Event?", 4.

milieu, the excrescence—or, the out-growth—of a vegetal life articulates a prefiguration of any finitude or terminal outcome on the one hand, and a "continuous appearing, forever indebted to and drawing on its 'soil.'"[29] According to Marder, "growth is never 'mere,' if only because the coming-outside-itself of a growing being, its exposure to the world, is an event, notably from the phenomenological point of view."[30] This relationality expresses an attunement to the changing milieus as vegetal temporality, as Marder notes, is a lived expectation of "the alterations and alternations of planetary time".[31] Being situational, embedded, contingent and unstable, the eventfulness as manifested in the vegetal mode of being introduces exceptionality and indeterminacy into the destiny of a growing entity.[32] Similar to the Chinese character *huā*, the notion of vegetal event, understood as a rooted continuity and an earthed dynamism in sync with the interwoven milieus, is itself an emerging *zào huà* (creation and transformation) of a being's openness to, and engagement with, the nuanced field of interconnectedness—more specifically, the field of affect wherein the affectings and beings-affected of an event's avatars contribute to the bloom of its eventfulness.

As the notion of vegetal event affords us a rhizomatic rootedness *with-in* the environment interlaced with differential agencies, the concept promises a fruitful rethinking of the Anthropocene by envisioning an attunement to—or in Marder's term in *The Philosopher's Plant*, "flowering-with"[33]—myriad earthly rhythms with plants. The difficulty, however, is how this affective attunement to the vegetal plant becomes possible given our tendency to *voice for* other nonhuman beings in accordance with human *logos*. On this point, Luce Irigaray has elucidated on the significance of silence in her *Through Vegetal Being* co-written with Marder. For Irigaray, silence is the key to "a being-with, without domination or subjection", itself being "the first dwelling for coexisting in difference."[34] Similarly for Marder, the coexistence in difference depends on "a balance between listening and noninterference, learning from the other or from others and respecting her, his, its, or their incommunicable otherness."[35] In other words, silence renders possible—or, eventuates—an affective attuning to the vegetal event. Interestingly, influenced by both Daoism and Chan

29 Ibid., 6.
30 Ibid., 7.
31 Ibid., 9.
32 Ibid., 11–2.
33 Marder, *The Philosopher's Plant*, 218.
34 Luce Irigaray and Michael Marder, *Through Vegetal Being: Two Philosophical Perspectives* (New York: Columbia University Press, 2016), 50.
35 Ibid., 114.

(Zen) Buddhism, the state of stillness and quietude is ubiquitous in Chinese landscape poems by the poet Wang Wei and his contemporaries. As the Chinese poet and critic Yip Wai-lim argues in *Diffusion of Distances*, "[t]he 'voices' one hears are those one hears in absolute silence voices from unspeaking, self-generating, self-conditioning nature outside the world of language."[36] Indeed, silence voices—it enunciates a harmonious interrhythm among differential life forms, through which we attune ourselves to the plant-voices.

3 Scatterings of the Magnolia: Emptiness and the Silence of Vegetal Event

In his introduction to *Laughing Lost in the Mountains*, Willis Barnstone acutely notes that Wang Wei (699–759 AD), who was dubbed "the Buddha of the Poets," has the genius of profound simplicity: "In a voice, a particle, one bell in the mountain, or a shuddering leaf, he sees, as William Blake saw in his 'grain of sand,' a universe."[37] In Wang Wei's poetry, silence is manifested as the Daoist attainment of "the unbroken ecstasy of unity"[38]—which is achieved by *wuwei* 無為, the non-action—and the realization of the Buddhist notion of *sunyata*, emptiness, which in Chinese is *kong* 空. By portraying things *in* silence and *through* silence, Wang Wei enables the non-human phenomena to enunciate their agencies in lacunae—a veiled field of indeterminacy like the semi-continuous line of mist in Chinese paintings[39] that renders their ontological voices ungraspable and inexhaustive, wherein a genuine counter with the nonhumans becomes possible. As highlighted by Yip, Wang Wei's poems are non-metaphoric and non-symbolic: "The objects presented are nothing more than the objects themselves. The poet does not step in, but, rather, he allows the scenery to speak and act itself out. It is as if the poet has become the objects themselves."[40]

---

36 Yip Wai-lim, *Diffusion of Distances: Dialogues between Chinese and Western Poetics* (Berkeley: University of California Press, 1993), 113.

37 Willis Barnstone, "The Ecstasy of Stillness," in *Laughing Lost in the Mountains: Poems of Wang Wei*, eds. Tony Barnstone, Willis Barnstone and Xu Haixin (Hanover: University Press of New England, 1991), xci–xvii.

38 Ibid., xlviii.

39 It would be interesting to make a note here that Wang Wei was not just a poet but he also excelled in music and landscape painting. Wang Wei's poems are renowned of being imbued with pastoral beauty, and his paintings repleted with poetics (詩中有畫，畫中有詩).

40 Yip, *Diffusion of Distances*, 72.

Wang Wei's "Magnolia Village"《辛夷塢》is precisely such a manifestation of an object-oriented poetics that capacitates the vegetal beings to enunciate their agential continuities with the vibrant milieus in an eventful manner:

High on tree-tips, the hibiscus	木末芙蓉花，
In the mountain sets forth red calyxes.	山中發紅萼。
By the door in the ravine, quiet. No man.	澗戶寂無人，
It blooms and falls, blooms and falls.[41]	紛紛開且落。

This poem adumbrates the blooming and falling of the magnolia flowers through a sharp contrast between the vibrant reddening of the calyxes in the first two lines, and the quiet scatterings of petals in the last couplet; in short, the phenomenal form of the floral beings and emptiness that resides in their florescence and senescence.

The vegetal eventfulness of this poem is delineated in a twofold manner: in the reddening of the magnolia calyxes that expresses their embodied embeddedness *with-in* the milieu, and the quietude that expects the unsubstitutable singularity of each of the blooming and falling. The first two lines situate the magnolia flower, which is blooming and reddening like a hibiscus, by a zoom-in perspective focusing on the tree-tips, which is followed by a panoramic view of the mountain for sketching a lively scenery of the magnolia village agentially crimsoning and embellishing the green landscape. On the one hand, this reddening of the magnolia calyxes could be read as a plant-voice that articulates the continuity and contiguity of the floral beings to the tree and the mountain. As Marder constantly reiterates, language does not necessarily unfold in words; "gestures and living bodies speak as well, even in the way they inhabit places."[42] In other words, the reddening of the calyxes expresses an innate dynamism of the plants and the vegetal proliferation of "appear[ing] better, more fully" whilst "giving something new to sight and to all the other senses" and becoming ample "in concert with the environment".[43] On the other, the character *fa* 發 denotes the eventful *happening* (as in *fasheng* 發生) of how the magnolia flowers crimson the surrounding by *emanating* (as in *sanfa* 散發) vibrancy and vitality in their coming-into-appearance (as in *shengfa* 生發) and

41 Yip Wai-lim, *Chinese Poetry: Major Modes and Genres* (Berkeley, Los Angeles & London: University of California Press, 1976), 310. Two amendments have been made on Yip's translation. The first is the substitution of the transliterated title "Hsin-I Village" with a literal one "Magnolia Village." The second is the translation of "澗戶," which Yip translates as "A home by a stream". Here the term denotes a ravine that ensembles a door, and therefore it should be translated as "By the door in the ravine".

42 Irigaray and Marder, *Through Vegetal Being*, 113.

43 Marder, "What's Planted in the Event?", 4–5.

self-exposure to the environment. In and of itself, the blossom of magnolia flowers—the reddening of the calyxes—is an eventful resonance between the vegetal vitality and the changing context, namely, the seasonal rhythm that conditions the florescence.

Unlike the first two lines which depict the phenomenal form of the reddening magnolias, the last two lines elicit a sense of emptiness and quietude from the fallings of the plants. Having adumbrated the vibrancy of the crimsoning magnolias, the third line, interestingly and surprisingly, juxtaposes the lively scenery to a desolate ravine, contrasting proliferation and vitality with solitude and emptiness. As noted by Tao Wen-peng, this contrast is to evoke the sense of Zen 禪 through illuminating on the empty nature behind all phenomena—that the flowers fall self-so-ly, without any sorrow or lament.[44] The Buddhist notion of emptiness is crucial to appreciate the vegetal event in the last line of the poem. According to the Chinese philosopher Mou Zongsan, Buddhism strives to remove the notion of permanency "by advocating that '[phenomena are] emptied of substance' (性空), 'all dharmas have no self-nature' (諸法無自性), and 'all acts have no permanency, all dharmas have no substantial-self.' (諸行無常，諸法無我)"[45] In her *A Philosophy of Emptiness*, Gay Watson explains that the notion of emptiness "is the interdependence of all phenomena: to be empty is to be dependently originated."[46] In this sense, the magnolias—their bloomings and fallings—are emergent processes eventuated by the interaction and interdependence of their rooted relationalities to other elements.

The eventfulness of the last line of the poem perhaps resides in the rhythmic quality of the Chinese *die-zi* 疊字, or literally "folded characters," *fenfen* 紛紛. Here, credits should be given to Yip's illumination on this repetitive quality of *fenfen* in a way that the repetitions "blooms and falls" beautifully capture the eventful scenery of blooming and falling with respect to the vegetal life cycle. The outpouring extravagance of scattering magnolias, which fall ones after ones, instills difference—or in Marder's terms, alteration and alternation—into the individuated rhythm of each flower: "Each time anew the earth regenerates; every plant that sprouts is unique; each generation—unrepeatable."[47] Hence, the condensation of a vegetal life—blooming and falling—within this single line amplifies the quietude of the aforementioned reddening of calices

44 Tao Wen-peng, *A Bright Moon Shining the Poet Buddha in Pine Trees: An Appreciation of Wang Wei's Poetry* (Taipei: The Way International Cultural Limited, 2003), 60–1. Translation mine.

45 Mou Zongsan, *Nineteen Lectures on Chinese Philosophy: A Brief Outline of Chinese Philosophy and the Issues It Entails*, trans. Esther C. Su (Foundation for the Study of Chinese Philosophy and Culture, 2015), 262. Bracketed words by the translator.

46 Gay Watson, *A Philosophy of Emptiness* (London: Reaktion Books, 2014), 31.

47 Irigaray and Marder, *Through Vegetal Being*, 145.

by returning its vibrancy to emptiness, which in turn accentuates the unrepeatability and irreplaceable singularity of each magnolia—its florescence and senescence as an emergent particularity eventuated by a specific and unrepeatable interrelation to the interlaced milieu.

4 Fallings of the Cassia: an Attunement to the Vegetal Silence through Non-action

Whereas silence in Wang Wei's "Magnolia Village" voices the irreplaceable singularity of the floral beings, his "Bird-Sing Stream"《鳥鳴澗》manifests an affective attunement to the profound silence of the plant world by delineating a nuanced vagary of falling cassias:

Man at leisure. Cassia flowers fall.	人閒桂花落，
Quiet night. Spring mountain is empty.	夜靜春山空。
Moon rises. Startles—a mountain bird.	月出驚山鳥，
It sings at times in the spring stream.[48]	時鳴春澗中。

This poem limns the emergence of a subjectless subjectivity—a distributed agency of the event that synthesizes and harmonizes the cosmic flow of energy manifested differentially in myriad phenomena. According to Wong, this poem is "a concrescence of vibration and rhythm, iteration and reiteration without any obstruction of the heart or hanging on to forms."[49] The first line unleashes a juxtaposition of movements between the "man at leisure" and the falling cassias, with the two rhythms affecting and being affected by each other in an interplay of agencies that prepares the eventful scenario of the night *quieting* itself in the second line. Here, noting the syntactic flexibility of Chinese poetry, Wong argues that the translation of "night quiets" is more faithful to liberate the character *jing* 靜 as a verb that enriches the eventful quality of the *quieting* of the night.[50] Later in the third line, the poet further adumbrates this profound tranquility in which even the rising of the moon "startles" the mountain birds, evoking echoes and resonations among the falling cassias, the rising moon, and the singing of the birds in the spring stream as depicted in the last line.

48 Yip, *Chinese Poetry*, 306.

49 Wong Kin-yuen, "Buddhist Consciousness, Deleuzian Ecoethics, and the Case with Wang Wei's Poetry," *The SSEASR Journal* (2008): 46.

50 Ibid.

Albeit the vegetal imagery is present only in the first line, it should be highlighted how the falling of the cassias eventuates the perceptual threshold of attuning to the profound tranquility, from which the inter-rhythm of disparate phenomena emerges and unveils itself to the poet. As Yip argues, the acumen of this poem resides in the unfastening of mind that frees the conscious self to the non-projective and non-interpretive realm wherein a million things appear as they are: "Phenomena do not need 'I' to have their existences; they all have their own inner lines, activities, and rhythms to affirm their authenticity as things."[51] This spiritual state is attained through the Daoist practice of *wuwei*, literally the non-action, which, according to Mou, designates "tranquilness resulted from the voidful unity of the mind."[52] Lest it be misunderstood, *wuwei* is not any passive attitude of doing no actions; as Roger Ames and David Hall clarify in their philosophical translation of *Daodejing*, *wuwei* refers to "[a]ctions uncompromised by stored knowledge or ingrained habits", themselves being spontaneous, self-so-ing and nonassertive.[53] It is precisely through cultivating a leisure mind, unfastened from the conscious thinking self, that the falling of cassia flowers discloses itself to the wondrous mind of the poet in a detour towards multitudinous entities of differential rhythms, with tranquility of this poetic scenery voicing the affective vitality of infinite nonhuman agencies.

As the threshold to such a distributed subjectivity attuning to differential rhythms, the eventfulness imbued in the falling cassias synthesizes the affective relationships among disparate phenomena, enfolding *a* world from within. According to Tao, the falling of the cassias is key to the unfolding of the interlaced movements of differential phenomena in the poem: "It is the leisure state of mind that eventuates the knowing of the falling flowers; and the falling flowers eventuate the sensing of the quietude of the night, which at last eventuates the realizing of the mountain being empty."[54] This eventful continuity unfurls a singular imbricatedness among seemingly dissonant phenomena, evoking the inseparability of all things as manifested in the notion of *he* 和—harmony—in chapter 42 of *Daodejing*: "[Everything] blends these vital energies (*qi*) together to make them harmonious (*he*). 沖氣以為和"[55] This rings a bell with what Emanuele Coccia has in mind about "reciprocal imbrication",

51 Yip, *Diffusion of Distances*, 80.

52 Mou, *Nineteen Lectures*, 95.

53 Roger T. Ames and David L. Hall, *Daodejing: "Making This Life Significant": A Philosophical Translation* (New York: Ballantine Books, 2003), 39.

54 Tao, *A Bright Moon*, 27. Translation mine.

55 Ames and Hall, *Daodejing*, 143. Bracketed word mine.

according to which "everything is in everything (*pan en panti*)."[56] As an earthed synthesizer of its milieu, the plant life does contain within itself *a* world: the cassia flowers are themselves prisms refracting rhythms of the poet at leisure, the rising moon, and the startled mountain birds. As Wong notes the ideogrammatic design of the character *lao* 落 (down-going), the character contains within itself "a landscape crowned by a grass-flower radical, with water on the lower left and a foot entering a cave on the right."[57] Etymologically, the "falling" is itself a concert with differential elements. From within the profound tranquility emanated from the falling cassias, an inter-rhythm emerges to manifest to the wondrous mind of the poet the imbricated reciprocity and the affective resonance among all phenomena. On this point, Marder reiterates that "plants, along with other non-human beings, do not have a world" but only "a different relation of vegetal beings to their environment": "Whenever human beings encounter plants, two or more worlds (and temporalities) insect: to accept this axiom is already to let plants maintain their otherness, respecting the uniqueness of their existence."[58] In the case of "Bird-Sing Stream," the plant-world emerges from a harmonized intersection of myriad imbricated worlds, itself being an inter-rhythm articulating our embodied embeddedness with-in the shared environment as evinced in the notion of *xiangyun* 相蘊 (mutual imbricatedness) in *Zhuangzi*. Through attuning himself to the rhythm of the cassia flowers, the subjectivity of the poet becomes distributed, following the prismatic temporalities and agencies imbricated within the plants, affecting and being affected by the eventfulness enfolded within the *lao* (falling) in the genuine encounter with its profound tranquility.

5 Flowering-with the Stream Flower: Silence as an Eventful Encounter with Plants

For the purpose of adumbrating a rigorous scenario of an affective attunement to the vegetal event together with the notion of silence advocated in Buddhism and Daoism, we may resort to the poem "In Search of Taoist Priest Changshan's Cabin in Nanxi" 《尋南溪常山道人隱居》 by Liu Changqing 劉長卿 (709–785 AD), in which the poet forgets words in his encounter with stream flowers, with his wondrous mind following the Zen affect in tranquility:

56 Emanuele Coccia, *The Life of Plants: A Metaphysics of Mixture*, trans. Dylan J. Montanari (Polity, 2019), 66–7.
57 Wong, "As the Tree Greens," 300.
58 Marder, *Plant-Thinking*, 8.

一路經行處，莓苔見履痕。
白雲依靜渚，春草閉閑門。
過雨看松色，隨山到水源。
溪花與禪意，相對亦忘言。

Along the path taken, footprints on moss and lichen.
White clouds leaning on the still sandbar, spring grass encloses the leisure door.
After rain, admiring the tint of pines, following the mountain to the source of stream.
Stream flowers and the Zen affect, facing each other, forgetting words as well.[59]

Vegetal agencies are ubiquitous throughout the whole poem. As explained by the title, the poem is about the search for a Daoist priest Chanshan, which unexpectedly turns out to be the poet's wondrous immersion within the plantation world, with "moss and lichen" being his companions on his path taken at the beginning of the journey. As the poem continues, it is revealed that the poet's search is in vain as indicated by the enclosure of "the leisure door" by the "spring grass." Here, the spring grass is granted agency as the subject of the fourth line who "encloses the leisure door." Interestingly, vegetal agencies hidden behind the ideogrammatic design of the character *xian* 閑 (leisure), which pictorializes a wood (木)—or a tree—inside a door (門), suggest an appeal to the poet to expect the plant world leisurely and wondrously. Since then, the second half of the poem unravels a vibrant botanical world of interwoven non-human agencies: as depicted in the fifth and sixth lines, the poet encounters the lively milieus of pines and waters, flowers and streams. Following the Zen affect emanated from the plant life, as declared in the last couplet, the poet forgets words eventually, himself being mesmerized completely in his encounter with the plants.

The eventfulness of the poet's attunement to the silence of the plant life resides in the subtlety of the character *xiang* 相, which, with its ideogrammatic arrangement of a wood (木) on the left and an eye (目) on the right, manifests

59 My translation of the last couplet refers to Wong Kin-yuen's essay "The Treeing-of-tree through Affective Attunement." (53) It is interesting to note here that the translation of *yi* as affect, which denotes a felt resonance exceeding the grasp of human language, resonates much with what the Sung poet and poetry theorist Yan Yu 嚴羽 has eloquently articulated in *Canglang Poetry's Talk*《滄浪詩話》: "there is an end to speech, but no end to meanings (*yi*). 言有盡而意無窮" It is therefore my preference to retain the philosophical concept of affect in the translation.

a mutual expectance between humans and plants. According to Wong, *xiang* designates "an act of synchronizing, mutually transforming, exchanging and matching with each other between two entities."[60] Through a mutual facing of each other (相對), the poet and the stream flowers will the event of *wùhuà*—the mutual transformation of things—wherein the persona unfastens the conceptual awareness of self and liberates his subjectivity to the interlaced milieu of vegetal affects. Here, the translation of the Chinese character 意 *yi* as affect deserves an elaborate explanation. As well noted by Wong, the Chinese character "has an ideogrammatic composite of a sound or an act of sounding (音) above the heart (心)."[61] With *yi*, which literally means "meaning," morphologically taken as the sound above the heart, the character denotes a *felt* quality of affect shared among different entities that exceeds human language. In the poem, it is precisely through the mutual expectance with plants in the event of deforming the concept of subjectivity as denoted in *xiang* that an affective resonation with the plant-voice is eventuated.

In such an event of the poet synchronizing, transforming and becoming with the stream flowers, silence plays a pivotal role in engendering the detour from a rigid notion of subjectivity to the emergence of a flowering-with the plant life. The notion of tranquility is particularly embraced by both Daoism and Buddhism. According to Mou, tranquility does not merely point to "rest" as opposed to "motion"; instead, as in Daoist philosophy, it connotes the spiritual state "that is full of infinite, wondrous functions (無限妙用)."[62] With this infinite and wondrous function of tranquility, the poet's perception of the vibrant agencies of plants is rendered possible, being capacious to respond to their expectancy without imposing his voice upon the flowers. It is therefore the emphasis on cultivating silence as a means of responding to such infinite potentialities in chapter 16 of *Daodejing*: "Extend your utmost emptiness as far as you can/ And do your best to preserve your [tranquility] (*jing*). 致虛極，守靜篤"[63] Likewise in the chapter "The Natural Course of Events" 〈天道〉 in *Zhuangzi*, silence is regarded as the ultimate virtue: "Emptiness, peacefulness, quietude and non-action are the yardsticks of the heaven and the earth and also the ultimate criteria of Tao and virtue. 夫虛靜恬淡寂漠無為者，天地之平而道

60 Wong, "Treeing-of-tree," 52.

61 Ibid., 44. Bracketed words mine.

62 Mou, *Nineteen Lectures*, 96.

63 Ames and Hall, *Daodejing*, 99. In Ames and Hall's translation, the character *jing* is originally rendered as "equilibrium" to the state of ongoing balance between the self and the environment through a continual adjustment. The original translation of by Ames and Hall is "equilibrium."

德之至"[64] Emptiness and tranquility, as Ames and Hall remark, capacitate a non-subjective perception of the world "as it is without imposing [the mind's] own presuppositions upon it".[65] The poet's forgetting of words manifests such emptiness and tranquility of mind that eventuates this genuine encounter with the flowers, wherein the plants enunciate their agencies and articulate the plant-voice in affects imbued in silence, wherein the poet flowers-with the stream flowers in affective resonance.

Marder elucidates the spirit of "flowering-with" with an allusion to the image of the Buddha's gaze at a flower. As he writes, the Buddha's gaze, in its unperturbed silence, "accompanies the growth of the flower without indifference, grows *with* it, and welcomes it in all its otherness".[66] The speechlessness depicted in the poem is crucial to initiate an eventful encounter with the stream flowers because the realm of silence is where the radical differences of plants are perceived without indifference, where the vegetal agencies reveal themselves in lacunae devoid of human *logos*. Suffice it to say, it is precisely through the poet's forgetting of words that plant agencies emerge to the fullest in the affective resonance. This fidelity to the vegetal silence echoes Marder's notion of "flowering-with": "finding one's conditions for flourishing in the other and dedicating oneself to the other's flourishing".[67] As in the Buddha's gaze and the poet's forgetting of words, the eventfulness of flowering-with blooms in silence, in lacunae that expect.

6 As Silence Voices: Thinking with Plants in the Anthropocene

Plant voices in silence, and its silence voices a radical way of being with-in the world which, in keeping with the Latin etymology *radicalis*, elucidates our relationalities "of or pertaining to roots"; in short, our rootedness in the rhythms of this vibrant more-than-human world. As this chapter argues, a plant-voice is *earthed* in a profound silence wherein its rootedness to the seasonal rhythm expresses its non-substitutable singularity; and *unearthed* itself to a wondrous mind as an inter-rhythm wherein its voice expresses and is expressed by other rhythmic bodies. The spirit of attuning to the silence of vegetal life is to cultivate a deformed mind of emptiness and tranquility that eventuates the event of flowering-with the vibrant plantation worlds.

64 *Zhuangzi*, 201.
65 Ames and Hall, *Daodejing*, 100.
66 Marder, *The Philosopher's Plant*, 214. Original emphasis.
67 Ibid., 222.

The primacy of thinking the Anthropocene with plants is to envision an alternative plant-thinking to resist the dominant and self-destructive discourses of anthropocentrism that validate the enslavement of vegetal life for instrumental needs. It is therefore the task of any plant-thinking to salvage the vegetal silence from oblivion, and to reimagine a mode of co-existence that respects the otherness of plants without being indifferent to their differences: in other words, to re-encounter plants, genuinely without any presuppositions. As Marder acutely notes, plant-thinking is "in the first place the promise and the name of an encounter, and therefore it may be read as an invitation to abandon the familiar terrain of human and humanist thought and to meet vegetal life, if not in the place where it is, then at least halfway."[68] As in my demonstration of how Chinese poets attune themselves to the vegetal silence, a genuine encounter with plants resides in willing the event in which human and plant beings enter a mutual imbricatedness that transcends the subject-object boundary: to mutually expect and be expected, affect and be affected.

In conclusion, I would like to invoke a parable from the chapter "Wandering in Absolute Freedom" 〈逍遙遊〉 in *Zhuangzi* to wrap up my arguments on how to will the event with plants in a non-anthropocentric and non-humanist manner that counters the Anthropocenic discourses. In the parable, Huizi criticizes the appearance of a tree called *shu* 樗, commenting that its gnarled trunks and twisted branches render it useless. On this point master Zhuang replies:

> Now you have a big tree, but are worried about its uselessness. Why don't you plant it in the land of nothingness, a wilderness where nothing grows and no one comes? There you may roam idly around it and sleep carefreely beneath it. No one will apply an axe to it or do any harm to it. Useless as it is, how can it ever come to harm?[69]

Particularly in this geological epoch of the Anthropocene, plant-thinking scrutinizes the anthropocentric depiction of plants as inferior and instrumental objects. Instead, it appeals to a genuine approach to plants. As in the aforementioned poems or in the parable above, it is a matter of willing the event through attuning oneself to plants, of affording a flowering-with that voices an imbricated silence.

68 Marder, *Plant-Thinking*, 10.
69 *Zhuangzi*, 13.

Bibliography

Ames, Roger T., and David L. Hall. *Daodejing: "Making This Life Significant": A Philosophical Translation*. New York: Ballantine Books, 2003.

Barnstone, Willis. "The Ecstasy of Stillness." In *Laughing Lost in the Mountains: Poems of Wang Wei*, edited by Tony Barnstone, Willis Barnstone and Xu Haixin, XV–LXVII. Hanover: University Press of New England, 1991.

Coccia, Emanuele. *The Life of Plants: A Metaphysics of Mixture*. Translated by Dylan J. Montanari. Polity, 2019.

Gu, Jian-ping. *A Graphic Compendium of Chinese Characters*.《漢字圖解字典》Shanghai: Orient Press Center, 2008.

Harper, Donald. "Flowers in T'ang Poetry: Pomegranate, Sea Pomegranate, and Mountain Pomegranate." *Journal of the American Oriental Society* 106, no. 1 (1986): 139–53.

Irigaray, Luce, and Michael Marder. *Through Vegetal Being: Two Philosophical Perspectives*. New York: Columbia University Press, 2016.

Laist, Randy. "Introduction." In *Plants and Literature: Essays in Critical Plant Studies*, edited by Randy Laist, 9–17. Amsterdam & New York: Rodopi, 2013.

Marder, Michael. *Plant-Thinking: A Philosophy of Vegetal Life*. New York: Columbia University Press, 2013.

Marder, Michael. *The Philosopher's Plant: An Intellectual Herbarium*. New York: Columbia University Press, 2014.

Marder, Michael. "What's Planted in the Event? On the Secret Life of a Philosophical Concept." In *The Green Thread: Dialogues with the Vegetal World*, edited by Patrícia Vieira, Monica Gagliano and John Ryan, 3–17. Lanham, Boulder, New York & London: Lexington Book, 2016.

Massumi, Brian. *Politics of Affect*. Cambridge: Polity, 2015.

Mou, Zongsan. *Nineteen Lectures on Chinese Philosophy: A Brief Outline of Chinese Philosophy and the Issues It Entails*. Translated by Esther C. Su. Foundation for the Study of Chinese Philosophy and Culture, 2015.

Pan, Fuh-jiunn. *Botany in Chinese Literature*.《中國文學植物學》Taipei: Owl Book, 2011.

Pan, Fuh-jiunn. *Botany in Tang Poetry*.《全唐詩植物學》Taipei: Owl Book, 2018.

Ryan, John Charles. *Posthuman Plants: Rethinking the Vegetal through Culture, Art, and Poetry*. Illinois: Common Ground Publishing, 2015.

Ryan, John Charles. *Plants in Contemporary Poetry: Ecocriticism and the Botanical Imagination*. New York & London: Routledge, 2017.

Seigworth, Gregory J., and Melissa Gregg. "An Inventory of Shimmers." In *The Affect Theory Reader*, edited by Melissa Gregg and Gregory J. Seigworth, 1–25. Durham & London: Duke University Press, 2010.

Spinoza, Benedict de. *A Spinoza Reader: The Ethics and Other Works*. Translated by Edwin Curley. Princeton: Princeton University Press, 1994.

Tao, Wen-peng. *A Bright Moon Shining the Poet Buddha in Pine Trees: An Appreciation of Wang Wei's Poetry*.《明月松間照詩佛：王維作品賞析》Taipei: The Way International Cultural Limited, 2003.

Tong, Christopher K. "Nonhuman Poetics (By Way of Wang Guowei)" In *Ecocriticism and Chinese Literature: Imagined Landscapes and Real Lived Spaces*, edited by Riccardo Moratto, Micoletta Pesaro and Di-kai Chao, 38–54. London & New York: Routledge, 2022.

Watson, Gay. *A Philosophy of Emptiness*. London: Reaktion Books, 2014.

Wong, Kin-yuen. "As the Tree Greens: Deleuze's Form-Event Assemblage and Chinese Ideograms in a Biosemiotic Ecosystem." *Deleuze and Guattari Studies* 15, no. 2 (2021): 285–317.

Wong, Kin-yuen. "Buddhist Consciousness, Deleuzian Ecoethics, and the Case with Wang Wei's Poetry." *The SSEASR Journal* (2008): 37–56.

Wong, Kin-yuen. "The Treeing-of-tree through Affective Attunement: Biosemiotics and Chinese Ideograms as an Ecosystem." *Ecolinguística: Revista Brasileira de Ecologia e Linguagem* 6, no. 2 (2020): 32–61.

Yip, Wai-lim. *Chinese Poetry: Major Modes and Genres*. Berkeley, Los Angeles & London: University of California Press, 1976.

Yip, Wai-lim. *Diffusion of Distances: Dialogues between Chinese and Western Poetics*. Berkeley: University of California Press, 1993.

Zhuangzi. Translated by Wang Rongpei. Hunan: Hunan People's Publishing House, 1999.

PART 3

Human-Plant and Transfiguration

CHAPTER 9

Vegetal Apophasis: Reclaiming Dark Intimacies and Un/knowing with Plants

Clara Soudan

1 Encountering Intimate Aliens

In *Plant Thinking*, the philosopher Michael Marder claims that "an encounter with plants awaits us." This encounter would draw near to an untranslatable *desencontro*, "a narrowly missed meeting, a crossing of paths that was about to happen but ended up not taking place, or to an encounter that is too improbable and was never meant to happen, or, again, to a divergence of two or more (usually human) beings, each of them existing on her or his own wavelength" (2013, 13). About to delve into a philosophy of vegetal life, Marder thus heralds that the promise of our encounter with plants might be somehow deluding, disorientating, plunging us into a thick cloud of unknowing. Plants, he suggests, are *resistant* to human thought, their ontic efflorescence overflowing our attempts to conceptualize them. In Richard Powers' novel *The Overstory*, trees weave and whisper a world overflowing our cognition, inaccessible to our regimes of thought. A cryptic arboreal kingdom manifests itself to the human protagonists as a dark epiphany, addressing them, summoning them to listen to things said in "words before words":

> All the ways you imagine us—bewitched mangroves up on stilts, a nutmeg's inverted spade, gnarled baja elephant trunks, the straight-up missile of a sal—are always amputations. Your kind never sees us whole. You miss the half of it, and more. There's always as much belowground as above. (Powers 2019, 4)

Throughout his novel, Powers depicts vegetal life as an alien, ancient terrestrial presence which eludes humans all the while deeply, intimately inhabiting them. As anthropologist Natasha Myers beautifully reminds us, "plants breathe us into being", their exhale is the condition of possibility of our inhale (Myers 2020). In an article entitled "The Intelligent Plant", Michael Pollan recounts a conversation he had with the botanist Stefano Mancuso. The latter claimed that, if we could begin to understand plants on their own terms, "it would be like being

 | DOI:10.1163/9789004721777_011

in contact with an alien culture"—a benevolent and knowledgeable culture with their own modes of sensing, knowing and communicating (Pollan 2013). Bathed in this light, plants appear as alien, largely unknown kins which precede us and whose photosynthetic powers and "terrestrial magic" compose a breathable world allowing us animal creatures to roam the earth (Myers 2021). Against the concept of Anthropocene which eclipses much of the agency of more-than-humans, Myers' Planthroposcene proposes "to change the terms of encounter, to make allies with these green beings" which inhabit, nourish and sustain us (Myers, 2017, 4). This "aspirational episteme" rises as a call, stemming from a reclaimed encounter with the vegetal realm, to decolonize the way we feel and think. Echoing the emerging field of plant cognition and its invitation to acknowledge the cognitive abilities of plants including perception, learning, memory and consciousness, the Planthroposcene challenges us to regard plants as protagonists in the multi-species processes of terraforming the earth. The figure of the planthropos thus centers our terrestrial hybridities and the possibilities of world re/composing alliances with plants in the hope of seeding other sympoietic worldings. To initiate these new alliances beyond human exceptionalism and anthropocentrism, Myers suggests that we "reach towards the plants as if they are beings worthy of address"—thus overcoming what biologists James Wandersee and Elizabeth Schussler call our "plant blindness" (Myers 2020, Wandersee and Schussler 1999). Cultivating a posture of attentiveness toward these green aliens who breathe us into life might entail envisioning abilities to feel, think and communicate not as human prerogatives, but as a dimension of life which as such can also be found in vegetal life forms (Kohn 2013). Encountering plants, addressing them, listening to them, acknowledging their own modes of existence, in modernist terms, implies an appreciation of their own kind of subjectivity. In the perspective of a Levinassian ethics of alterity, it requires opening ourselves to their unreachable, unsubsumable and inexhaustible otherness. Awakening to our radical entanglement with vegetal lives thus summons us to attend to a radical intimacy with these terrestrial aliens and open ourselves to the presence of more-than-humans with and within us. Timothy Morton's dark ecological approach embraces such radical intimacy with radical strangers, claiming that the awareness of our entanglements with more-than-humans discloses an intimacy "full of ambiguity and darkness." Their concept of *strange strangers* articulates the uncanniness of an ecological consciousness that erodes our traditional sense of self by dissolving boundaries between culture and nature, organism and environment, inside and outside, self and other (Morton 2010, 100).

What draws my attention here is the conflation of a radical otherness with a no less radical intimacy in our relationship to vegetal life. Breaking open our ontological, epistemological and ethical regimes of thought, the *des-encontro* with the opacity of vegetal beings—these obscure "beings of light and air" as Powers calls them—promises a dive into our repressed intimacy with vegetal life—our being inhabited by plants. Marder's vegetal philosophy proposes to unfold an anamnestic work of remembrance summoning the recollection of a forgotten kinship, a symbiotic thinking entangled in vegetal life and vegetal modes of thinking. Across the anthropocentrism and humanism of our traditions of thought, we are invited to venture into the abyss insulating humans from vegetal lives and reclaim a sense of our radical relatedness and intimate entanglement with and within more-than-human beings. Overthrowing the naturalist alienation which permeates our engagement with the more-than-human world, collapsing boundaries between realms of life, Marder reformulates our human wanderings as "strange archives, surfaces of inscription for the vestiges of the inorganic world, of plant growth, and of animality—all of which survive and lead a clandestine afterlife in us, as us" (Marder 2013, 10). Doing so, he illuminates a radical porosity between life-forms, along with the intimate enmeshment and processes of compenetration in our entangled becomings. As he highlights the vestiges of the vegetal surviving in the human, Marder interestingly inverts the traditional idea according to which humanity imprints itself upon a virgin nature. Plants as well as humans, together with other more-than-human aliens of the earth, weave the fabric of our entangled lives along strange intimacies.

Marder's plant philosophy brings to light an aporia which infuses the present chapter, namely that our encounters with plants—in their industrial exploitation as well as in their philosophical neglect—tend to eclipse much of their *alterity*. The plant biologist and philosopher Sylvie Pouteau thus argues for the recognition of the ontological openness of plants (Pouteau, 2020). If, as Pouteau contends, representing the openness of vegetal beings is a major ethical as well as ecological issue of our time, then how may their open becomings be embraced? Can their burgeoning alterity only be rendered in a way that would not be alienating? What if the openness of plants pointed toward the edge of thought, language and becoming itself? How may we encounter what overflows or withdraws in others? These interrogations lead us to wonder about the possibility of a mode of thinking or sensing that would not reduce, enclose nor petrify, but rather embrace, cherish, and nurture the ever overflowing, radical openness of the uncanny beings which inhabit us. How to renegotiate the terms—literally the epistemic regime—of our encounters with plants so as the avoid the modernist pitfalls of totalization and enclosure? If plants appear to withdraw from our conceptual grasp, "the idea", writes Marder, "is to allow (them) to flourish on the edge or at the limit of phenomenality, of visibility, and, in some sense, of "the world"" (Marder 2013, 8). Following Marder, I reflect upon the epistemological conditions that would enable plants to exist on the edge of their appearing, beyond the veil of their manifestation, and beyond the scope of our cognition. How may we encounter plants without eclipsing their perpetual withdrawing, their radical overflowing?

Taking cues from Myers, who summons a "radical disruption to learn other ways to see, feel and know" (Myers 2021), I advocate cultivating modes of attention to our vegetal kins enriched by an apophatic sensitivity to their radical overflowing. In theological thought, the apophatic way characterizes a mode of mystical discourse embracing the ineffable and clearing a space for what cannot be named. It waves toward an ontological withdrawal, reminiscent of the way Martin Heidegger approached the unfolding of Being in the light of an active resistance in its manifestation—an inherent elusiveness, a presence that discloses itself through its very concealment (Heidegger 2003). Apophatic language discloses and embraces such withdrawal by drawing our attention to what eludes us and remains hidden. Apophasis may also find an echo in the process of Derridean deconstruction, in which "there is always a remainder that cannot be read, that must always remain *alien*" (Derrida et al., 2009, 4. I emphasize). Morton's critique of the modern concept of nature similarly suggests its failure to account for a perpetual withdrawal, an obscurity and the haunting presence of "aliens and ghosts" permeating our blurred selves all the while escaping our attempts to digest them (Morton 2010a, 100). Against ideologies of nature which tend to petrify, fetishize and reify complex processes of co-becoming, committed to the openness of the overflowing other, dark ecology "refuses to digest plants and animals and humans into ideal forms" (Morton 2010b, 269). As such, it resists what Friedrich Nietzsche depicts as the modern impulse to devour and digest all of the world: "Modern man understands how to digest many things, indeed almost everything—it is his kind of ambition" (Nietzsche 1997, 170). Dark ecology embraces the overflowing as a radical dimension of our being intimately entangled within the alienness of the world. Devoted to an opaque cloud of distance surrounding our entanglements, dark ecology pledges to clear a space for the intimate aliens who inhabit us. As attuned to Morton's dark ecology as it may be, the apophatic sensitivity endorsed in this chapter is first indebted to the analyses of feminist process theologian Catherine Keller. In *Cloud of the Impossible*, Keller draws together the theological tradition of apophasis with the uncanniness of our inextricable earthly entanglements, thereby embedding the cloudy lineage of unknowing in the materiality of our terrestrial enmeshment with human and more-than-human beings. Keller finds an apophatic echo in the way quantum physics unsays the laws of classical physics. She resorts to apophatic language to approach the confluence of the inextricability of our planetary entanglements with the radical indeterminacy or openness of our becomings. Drawing upon Keller's analyses, I wonder whether apophatic languages of unsaying might allow us to clear a space for a renewed encounter with plants, such as called for by Myers and Marder. Myers contends that it is imperative that we cede room and return some of our biopolitical control over to the plants, to create better conditions for them to grow and thus to "allow them to exercise their responsibilities to the Earth as they terraform it with their remarkable practices" (Myers 2020). Myers' vision to grow liveable worlds in the Planthroposcene thus includes the preservation of plants' rights to express their "full, unfettered powers" and "to freely articulate relations with their allies and express their fullest desires" (Myers 2021). Marder too pleads for an encounter with plants that would foster the proliferation of vegetal life, the amplification of vegetal voices and the preservation of their alterity, asserting that we have no "unconditional right of admission into the vegetal world" (PT9).

Our encounter with vegetal life demands that we embrace a radical intimacy with radical strangers while attending to the openness and alterity of our more-than-human kins. I suggest that reclaiming an intimacy with these terrestrial aliens invites us to cultivate an epistemic humility as well as a sensitivity toward more-than-human ways of un/knowing. As we shall see, plants summon an attention to processes eclipsed by dominant regimes of perception and thought. I submit that the obscurity of plants illuminates alternate modes of sensing and thinking characterized by an apophatic awareness to the presence and wisdom of that which escapes us. Reclaiming an intimacy with such opacity, I argue, is key to cultivate an aesthetical, epistemological and political sensitivity to the more than humans who inhabit and overflow us.

2 A Forbidden Intimacy

Registering an uncanny intricacy with the strangers populating our confounded selves, ecology is profoundly about intimacy—so reminds us Morton. How may we weave healing intimacies with plants across the porous abyss of dualism, from within the obscure mesh that holds us together? In "Vegetal Life and Onto-Sympathy", political philosopher Jane Bennett draws our attention to the way human agency is entwined in vegetal life. She observes a "sympathy" between animal and vegetal bodies as they multiply "strange attachments", contending that "instances of human-plant affinity abound once you become alert to them". Commenting on an encounter between David Thoreau, leaves and pine needles recounted in *Walden*, Bennett argues that the poet's profound intimacy with the vegetal and more-than-human beings around him gives witness to affinities between bodies, "a resonance that does not collapse into a simple sameness" whereby "complex bodies acknowledge a degree of overlap". She conjures up a movement of transubstantiation as she describes the kind of intimacy that occurs as we ingest plants and they penetrate our insides: "eating the leaves, berries, or seeds of plants, a new corporation is formed, in which veggies and flesh mutually transform" (Bennett 2020, 90, 93, 103, 92). Bennett draws upon Annemarie Mol's "I eat an apple", in which the ethnographer demonstrates how the mundane, daily act of eating and digesting manifests a dissolution of traditional agencies and subjectivities: "I have become (made out of) apple; while the apple is (a part of) me. *Transubstantiation*" (Mol 2008, 30). Attentive to the intimacies binding plants and insects together in processes of pollination, Carla Hustak and Natasha Myers aim to track "the very momentum through which organisms reach toward one another and involve themselves in one another's lives". They illuminate a movement of involution at play in the encounters between plants and insects—an enfolding, a curling inward at the heart of interspecies intimacies (Hustak and Myers, 2012, 96). Tuning in to such intimacies

> requires reading with our senses attuned to stories told in otherwise muted registers. Working athwart the reductive, mechanistic, and adaptationist logics that ground the ecological sciences, we offer a reading that amplifies accounts of the creative, improvisational, and fleeting practices through which plants and insects *involve* themselves in one another's lives. (Hustak and Myers, 2012, 77)

In a way reminiscent of Alfred North Whitehead's concept of prehensions, through which actualities perceive the world into becoming, Hustak and Myers claim that "the world is full of "propositions" waiting to be registered by *interested* bodies" (Hustak and Myers 2012, 105). The evolutionary ecologist and pioneer in the field of plant bioacoustics Monica Gagliano similarly argues that the world offers an endless flow of multiples opportunities *inviting* beings into action (Gagliano 2018, 53). What should we do of this radical echo, these uncanny affinities between vegetal and animal beings, and of all the invitations and possibilities of involution emanating from them? How, as Karen Barad beautifully puts it, may we be present and "alive to the possibilities of becoming" with plants (Barad 2007, 396)?

Like Anna Tsing or Merlin Sheldrake with fungi, Myers summons new alliances and intimacies with plants, calling us to "seed plant people conspiracies" in the work of growing livable worlds and learning to "breathe together more deeply, more fully, and in more transformative ways". Conspiring with plants and learning what pleasures them is key to what the anthropologist regards as a social justice project to allow all beings to breathe more deeply (Myers 2020). Michael Pollan's description of the decisive introduction of caffeine in the rise of industrial civilization is evocative of such plant-people conspiracies. In *This is Your Mind on Plants*, Pollan highlights the entangled destinies of those plant species who produce caffeine and human societies striving on a Protestant ethic of sobriety, rationalism and productivism. Channeling scientific studies which analyze the psychoactive compounds of caffeine as an evolutionary tool for plants to attract pollinators and disseminate, he claims that "its appeal to our species allowed it to circumnavigate the planet" (Pollan 2021, 97). Pollan's account of the ambiguous relationship between legal authorities and mind-altering substances—reminding us that caffeine is the most widely used psychoactive drug in the world—illuminates the demonization of certain plant-people intimacies. Psychoactive plants may alter the textures of our consciousness, expand our perception, overturn cognitive patterns and open new trajectories of thought. Psilocybin—a psychedelic compound produced by over two hundred species of fungi—even shows promising therapeutic potential in treating various conditions such as severe anxiety, treatment-resistant depressive disorder, PTSD or addiction (Sheldrake 2020, 116). In the case of opium poppies, to come back to plants, even the—once vernacular—*knowledge* of their potency is criminalized, not least the "propagation" of these flowers. "How can it be", Pollan wonders, "that such an inconsequential speck of seed could yield a fruit in my garden with the power to lift pain, alter consciousness, "make sadness go away""? Pollan draws our attention to a "miraculous correspondence" between the molecule produced by a flower and the physiological processes governing pleasure and pain in our brains, echoing what Bennett refers to as an ontological "sympathy" between vegetal and human bodies (Pollan 2021, 78, 79). Opium poppies manifest indeed a radical intimacy between plants and humans, one so deep as to elicit euphoria and relieve pain in those who venture to drink their floral brew. Pollan therefore wonders whether a plant's toxicity, addictiveness, pleasuring or mind-altering properties make it dangerous, suggesting that only altered states of consciousness and addictions enhancing work productivity or benefiting capitalist economy in one way or another, as the opioid crisis tragically indicates, are legitimate. Our relationship with mind-altering plants is an ancient one. What justifies the demonization of this intimacy? As Pollan reminds us, plants altering the quality of our consciousness have historically been considered a threat to institutional power—including coffee or fermented apples (Pollan 2021, 110). Our relationship to plants is also policed in the repression of ancestral indigenous knowledge about herbal medicines and the prosecution of botanical drugs considered a threat to the pharmaceutical industry—marijuana and peyote are thus classified as schedule I drugs alongside heroin. Recovering forbidden knowledge of plants, along with the power that lies in such knowledge, provides a way of reclaiming flourishing possibilities of intimacy with our vegetal kins. Myers suggests that our relationship to the vegetal world has been severed by the legacies of a colonial order enforcing scientific rationalism and discarding our entanglement in multi-species intimacies. The prohibition of anthropomorphism or animism in scientific discourses thus appears central to an epistemological apparatus preventing the recognition of more-than-human personhood (Myers 2020). At odds with a dualistic structure of thought which postulates an unbridgeable cliff between humans and more-than-humans, Myers observes a metamorphic intimacy at play in scientific practice, whereby practitioners manifest an openness to be transformed by the plants they study. The anthropologist thus describes the "affective entanglements of inquiry" along with the "near-shamanic, shape-shifting, and intimate dimensions of scientific practice" (Myers 2017). What if we considered the unpredictable agency, or even the sovereignty of more-than-humans in these intimate encounters? The botanist and member of the Citizen Potawatomi Nation Robin Wall Kimmerer thus approaches plants as teachers, regarding this stance as a fundamental assumption of indigenous environmental philosophy. Kimmerer mentions indigenous cultures' "sophisticated research protocols ... for learning from the plants", supported by the knowledge she gained from her own "experiences of intense focus and attention with plants" (Shechet 2019) (Kimmerer 2013). Michael Pollan relates the comforting intimacy he felt with plants while on a trip with *Psilocybe azurescens, also called "the blue angel"*: "I could feel their curiosity and what I was certain was an attitude of utter benevolence toward me and my kind" (Pollan 2018). The mycologist Paul Stamets submits that mushrooms attempt to communicate biochemically with humans through their hallucinogenic properties—a thought echoed by the biologist Merlin Sheldrake who, during a fascinating conversation with philosopher and magician David Abram, wonders whether the writing of his book *Entangled Life* had been prompted by "fungal angels". Fungi, he claims, "exhilarate everything according to their needs, wants, ends" and might as such have infected him with the intention of spreading their voices to humans (Pollan 2018) (Abram and Sheldrake 2021). In *Thus Spoke the Plant*, evolutionary ecologist Monica Gagliano claims that her research on plant cognition was prompted by a walk in her garden during which plants suggested her to study them, sometimes even giving her precise directions on how to conduct her experiments (Gagliano 2018). These various experiences of close encounters with plants describe an intimate process through which a plant's healing knowledge penetrates us, reaching us over unknown channels. Circumventing human exceptionalism to include multi-species intimacies might lead us to consider learning opportunities, sources of knowledge and ways of un/knowing that would otherwise be ignored. Why would such experiences be dismissed? Wherein lies the subversiveness of plant knowledge—understood as the human knowledge about plants as well as the knowledge detained and sometimes shared by vegetal beings? Myers mentions the "power of aligning ourselves with plants", of cultivating solidarity projects with them, reminding us that "the tiniest bit of soil" suffices to initiate the work of transformation and activate healing processes for local ecologies. Plants offer an abundance of ways to ease the "powerful grip that capitalism has on our imagination" and initiate a process of decolonization of our minds and lands (Myers 2020). There is hope in the recovered intimacy with our vegetal kins, enabled by a posture of reverence and openness toward their therapeutical powers of healing, transformation and un/knowing. "Like all intimacy", writes Marder, "it will take place (largely) in the dark, respectful of the obscurity of vegetal life (Marder 2013, 181).

3 Un/thinking with Plants: In-betweenness, Openness, Becoming

What might a reclaimed intimacy with the obscurity of vegetal life disclose that present regimes of thought eclipse? As we contemplate ways in which plants might help us think and unthink, I embrace a diffractive thinking such as put forward by feminist philosopher and quantum physicist Karen Barad. Indebted both to a legacy of feminist theorizing of difference in the new materialist tradition and to the insights of quantum field theory, the notion of diffraction designates a critical practice which "troubles dichotomies" in a movement of "cutting together-apart" (Barad 2014, 168). Diffractive analysis is attentive to the flourishing of difference as an enfolded, entangled process, endorsing a relational approach to difference rather than a segregational one. Barad writes that differentiating "is not about radical separation, but on the contrary, about making connections and commitments" (184). A diffractive approach to the ways plants enable us to un/think and un/know remains therefore attuned to a process of epistemological proliferation, an opening of thought through which what makes us unthink simultaneously makes us think anew, and the unknowing prompted by plants offers alternative ways of knowing. If vegetal life illuminates the obsolescence of traditional metaphysics, as Marder demonstrates in *Plant Thinking*, and if, as Myers contends, we need a "radical disruption to learn other ways to see, feel and know", then approaching the deconstruction of metaphysics allowed by our encounter with plants as a diffraction implies a movement of involution, an enfolding rather than an overcoming of traditional patterns of thought in which we are already entangled (Myers 2021). As such, it echoes Marisol de la Cadena's wish to resist the "impulse to divide" and to "make alliances in and across divergence" (de la Cadena 2021, 256). A diffractive understanding of our philosophical encounter with plants thus embraces a potential efflorescence of thought, a germination from a fertile in-betweenness, an immanence birthing the world into entangled becomings, perpetually recomposing us.

Marder's project proposes to regenerate thinking through a renewed encounter with plants, to resume the process of philosophical thought as we stand in between collapsing and burgeoning worlds, on the edge of thought and language. *Plant Thinking* manifests an effort to re-embed thought in its relationship with plants and acknowledge the vegetal in us. What would plants teach us if philosophy's love of wisdom included their vegetal wisdom? In "Vertimus: Ten Theses on Becoming Plant", Marder claims that "guiding plants toward their most optimal becoming, caring for the soil and for what grows in it demands nothing less than overturning the status quo" (Marder 2020, my translation). He draws from the practice of turning over soil to argue that the processes and phenomena of vegetal life can inspire the deconstruction of Western metaphysics and its dualistic structure of thought. "Plants", he writes, "accomplish a living reversal of metaphysical values" (Marder 2013, 56). The parochial distinctions between self and world, subject and object—which "dogmatically separates us from what we aspire to know"—individual and community, being and becoming crumble indeed as we endeavour to think with plants. Thinking with plants entails thinking outside the totalizing categories of Western metaphysics: it encourages us to dwell in cracks, on the edge of thought, open to unpaved trajectories of becoming and pursuing the "atelic dispersion" of vegetal beings. Marder thus confers upon the figure of the plant, which "incarnates everything the metaphysical tradition has discarded as improper, superficial, inessential, purely exterior", the power to subvert and transvaluate metaphysical systems of thought. The tradition of metaphysics, he argues, is built upon a repression of vegetality: "resistant to idealization", proliferating in between rigid concepts, plants trouble, threaten and overflow traditional structures of thought (Marder 2013, 181, 90, 13).

One way they do so is through their ontological in-betweenness, which collapses traditional dichotomies and blurs boundaries between realms "for the place of the weed, much like that of existence itself, is precisely in between". Plants inhabit the in-between indeed: they mediate the earth and the sky, darkness and light, soil and air, the dead and the living—"caressing the dead with its roots and obtaining nourishment from them, the plant makes them live again" (Marder 2013, 90, 66). Dwelling simultaneously in different worlds, plants resurrect the dead into the living earth through the process of decomposition and breathe the sun into air through the process of photosynthesis. The liminal dwelling of plants is illustrated in Gilles Deleuze and Félix Guattari's rhizome, which has "neither beginning nor end, but always a middle from which it grows and from which it overspills" (Deleuze and Guattari, 1987, 21). Deleuze and Guattari do no less than summon a renewal of philosophical thought through a vegetal metaphor indicating that germination emerges from a space of in-betweenness, thereby subverting the metaphysical trope of the ground and the origin. Through them, the figure of the plant has become archetypal of post-metaphysical thought, inviting us to multiply nomadic connections in a movement of dispersion or dissemination. Marder explains that

> it is this middle place—not a fictitious inaccessible origin—that holds the promise of growth and proliferation, dispersed from the moment of its germination. Unlike the center, it is neither gathered into a unity nor oriented in a single direction; in its sheer materiality and organicity, the plant interferes with the metaphysical fixation on the One. (Marder 2013, 65)

Dissolving the idea of center and peripheries, rhizomatic thought draws our attention to a profusion of intersections, a diversification of trajectories. Plant thought sanctifies our radical entanglement in an inextricable, organic mesh of interconnectedness, thereby subverting hierarchies along with the philosophical obsession of separation which Marder regards as a vestige of the theologico-metaphysical tradition. Whereas the modern self is established in the myths of his autonomy and alienation from the otherness of the world, plants appear immersed in their surroundings, striving for alterity and thus weakening boundaries between self and world, as manifested in the "wood-wide-web". Through a cooperative subterranean network of fungi and bacteria called mycorrhizal networks, trees connect to each other, conveying warnings or delivering water and nutrients to their kins. If the rhizomatic wisdom of plants lies in their proliferating connections with a ubiquitous other, disclosing the world in all its interconnectedness, for Sylvie Pouteau, plants perform a type of subjectivity which dissolves boundaries between inside and outside (Pouteau 2020). This idea echoes Karen Barad's thought of a ubiquitous ontological touch, an intimate inhabitation, an involution of the stranger within whereby the world is composed (Barad 2018).

Marder identifies the restlessness of plants as another way through which vegetal life performs a deconstruction of metaphysics, which systemically disregards its vivacity. Attuned to the circular rhythm of the seasons, plants exhibit a constant regeneration, a perpetual growing and decaying. Their "ceaseless striving toward the other and becoming-other in growth and reproduction" resist metaphysical attempts to petrify the fluidity of vegetal beings (Marder 2013, 131). Manifesting a perpetual metamorphosis, open-ended, inexhaustible, proliferating, plants defy the metaphysical obsession for an immutable, static being:

> If incompletion means open-endedness, then vegetal growth fully satisfies this rendition of *ateles*, in that it knows neither an inherent end, nor a limit, nor a sense of measure and moderation; which is just another way of saying that it is monstrous and that *equally monstrous and unbounded is the thought germinating in it*. The life of a plant, metonymically associated with its growth (to say "violets grow in my garden" actually means "violets live in my garden"), is a pure proliferation *bereft of a sense of closure*. (Marder 2013, 24, my emphasis)

Marder draws our attention to the porosity of vegetal life with vegetal thought, suggesting that plants' unbounded dispersion permeates the thoughts germinating from our intimate encounters with them. Their ceaseless becoming thus appears as a becoming-with, a *worlding* in the new materialist sense of an open process of enmeshment, an active involvement in the material-semiotic events composing the world (Palmer and Hunter 2018). With the idea of sympoietic terran worldings, Donna Haraway reminds us that processes of world-composing and decomposing are always entangled in more-than-human, multispecies intimacies, rooted in the overflowing possibilities of our co-becomings (Haraway 2016). These worldings echo Myers' *involutionary encounters* whereby scientists "become with and alongside the living beings they study" (Myers 2017). Following Goethe, who claims that metamorphosis, or the process of becoming-other, is the essential feature of vegetal life, Marder goes so far as equating vegetality with becoming itself. Thinking with and alongside plants thus allows him to unfold a philosophy of endless becoming and embrace trajectories of un/becoming. "Becoming-plant", he claims, "we become becoming":

> If becoming is both a coming-to-being and a dying, a becoming and un-becoming, coming and going, coming and becoming, then it is both growth and decomposition, metamorphosis, or shape-shifting, occurring

> in between these concrete transitions. Which leads me to my first thesis: all becoming is a becoming-plant. (Marder 2020, my translation)

Committed to inexhaustible trajectories of un/becoming, plant thought yearns to preserve this ontological openness of vegetal life. Biologist Sylvie Pouteau thus engages plants as radically open beings, whose perpetual un/becoming manifest "the continuous unfolding, open-ended stream of life" (Pouteau 2018, 91). As "open beings", plants subvert the definition of a Newtonian object: they have no fix boundaries, no center, no inside nor outside. Ignoring boundaries between inside and outside, their ontological openness tends a ubiquitous liminality between beings, elements, realms of life. Vegetal becoming may thus be approached as a process of perpetual re-composition, an ingathering of the world in fluid intimacies. In "No Small Matter: Mushroom Clouds, Ecologies of Nothingness, and Strange Topologies of Spacetimemattering", Karen Barad emphasizes a radical openness haunting the heart of matter, a constitutive "void" full of "a desiring orientation toward being/becoming, flush with yearning and innumerable imaginings of what could be/might yet have been". Drawing upon Quantum Field Theory, they claim that "matter is spectral, haunted by all im/possible wanderings, an infinite multiplicity of histories present/absent in the indeterminacy of time-being", thus pointing to the presence of a germinating abyss in matter (Barad 2017, G112–G113). This quantum indeterminacy disrupts the metaphysics of substance—rooted, as Marder points out, in the repression of obscurity: "such repression of darkness severs the intellect, seeking pure light, from its roots swathed in obscurity" (Marder 2013, 177). In *Face of the Deep: A Theology of Becoming*, Catherine Keller thus discloses *tehomophobic* (fearful of the deep) tendencies in western Christian tradition, connecting the repression of feminized abysses to the desire to master chaos and control a fantasized unilateral origin. This theological loathing of the deep echoes Marder's diagnosis of a metaphysical repression of darkness resulting in the disavowal of vegetal life and earthly immanence:

> Western metaphysics commences, therefore, with the inversion of the earthly perspective of the plant, a deracination of human beings from their material foundations, their transplantation into the heavenly domain, and the correlative devaluation of the literal plant mired, with its roots, in the darkness of the earth as well as in non-conscious existence. (Marder 2013, 57)

Against the cultural repression of an impenetrable depth, unfathomable entanglements and an obscure openness, Keller articulates a *tehomic* theology of becoming which explores the hypothesis of a creation *ex profundis* (out of the depths) and offers a narrative embracing the darksome and chaotic complexity of the creative process. Un/thinking with plants awakens us to this fruitful darkness. Attending to their ontological openness, their liminal in-betweenness and the fluidity of their unbounded becomings invites us to subvert metaphysical and platonic structures of thought—drawn to the dazzling light of intelligible concepts. Philosopher Gianni Vattimo regards Marder's *Plant Thinking* as "entrenched in weak-thought" in so far as it fosters a "weakening of the strong structures of metaphysics" and endorses a posture of resistance against the violence of systematization and reduction (Marder 2013, XII). Vattimo's weak thought compels ways of thinking and knowing which do not aim toward an enclosure of thought but rather seek to unfold unpaved trajectories of becoming. The vegetal dissolution of metaphysical categories summons us to dive further into the darkness of plants, to divert from dominant regimes of knowledge and reflect upon what they conceal.

4 Vegetal Apophasis: Radical Unknowing to the Sound of Silence

The idea of vegetal apophasis suggests that vegetal life waves to modes of un/knowing withdrawn from our perception and comprehension. It acknowledges the profound obscurity of plants, their unfathomable strangeness and perpetual withdrawal. Even as we embrace radical intimacies with vegetal beings and immerse ourselves in the earthly entanglements that root us into more-than-human lives, we still find ourselves gasping on the edge of thought and language. Boundaries between inside and outside dissolve and, as we dwell in the liminal, our regimes of thought are overflowed by the inexhaustible. Therefore, to approach vegetal life, as Marder writes, "one must acquire a taste for the concealed and the withdrawn" (Marder 2013, 28). How to draw our attention to what eludes us? How may we attune ourselves to stories told in muted registers? How to address the intimate strangers who grace us with their darkness? Reclaiming our attention to such obscurity and to the modes of un/knowing detained by vegetal life entails cultivating an awareness for all that escapes present regimes of thought and knowledge. Natasha Myers calls for an ethic of unknowing in our relationship to plants, for a suspension of knowledge so that we may attend to their calling, rendered imperceptible by the spells we live under—hegemonic narratives, discursive regimes, systems of thought. Myers invites us to decolonize our imagination and open our sensorium to more-than-human sensibilities so that we might feel differently, sense with plants and attune ourselves to vegetal languages of unsaying. This suggests that we allow ourselves to be present to the unknown, opening ourselves to the teaching we may receive from others—other-than-human, vegetal beings. The mode of knowing Myers aspires to "unlearn radically" thus designates a predatory approach to knowledge and to more-than-humans as something that ought to be conquered, mastered, "dissected, extracted or parsed" (Myers 2020). This invitation to engage in alternative modes of perception to meet the world of plants echoes David Abram's call to deepen our sensorial awareness and synesthetic perception of the more-than-human world.

In the *Spell of the Sensuous*, the magician and philosopher diagnoses western philosophy's alienation from the sensory world, floating in a disembodied world of pure abstraction. Abram argues that our cultural disengagement from the world of sensory experience prevents us from responding to a more-than-human world that is always addressing us. To be responsive to our fellow earthlings and live in reciprocity with the myriad intimate strangers inhabiting with and within us, we must awaken to other forms of sentience and sensibility as well as to the expressiveness of the more-than-human world. Abram summons a reawakening of sensory faculties of perception which have lain dormant, obscured within western literate culture. Striving "to enter, ever so deeply, into the sensorial present", his approach surrenders alphabetic language and modes of thought to the expressive silence of the sensuous: "planting words, like seeds, under rocks and fallen logs—letting language take root, once again, in the earthen silence of shadow and bone and leaf" (Abram, 1996, 161–162). Abram thereby resumes phenomenologist Maurice Merleau-Ponty's project to return to a "world which precedes knowledge" and recover "the primacy of perception" in our engagement with the world (Merleau-Ponty, 1962). Embracing more-than-human intimacies implies indeed opening ourselves to the burgeoning invitations, stemming from the more-than-human world, to alter and expand the possibilities of our perception—thus relinquishing the fantasy of an all-encompassing language. Acknowledging more-than-human sentience and expressiveness is a movement of epistemological humility for it clears a space for other modes of knowing and saying to which we might have no access. Michael Pollan thus reports that "plants have evolved between fifteen and twenty distinct senses, including analogues of our five" (Pollan 2013). Abram's claim according to which everything not only senses but also speaks is supported by Serpil Oppermann's notion of storied matter. The material ecocritic contends that the entities with whom we interact are "alive, undeniably expressive" and "have their own stories to tell". That we only have a limited access to these stories is another matter and should not lead us to ignore the expressiveness of the more-than-human world. Beyond the "unbreakable" silence of vegetation, Marder similarly asserts that vegetal life expresses itself. The philosopher emphasizes the materiality of "vegetal self-expression" as plants release airborne chemicals or articulate themselves in space in a secret dance (Marder 2013, 74–75). Embracing the narrative agencies of more-than-humans enables us "not only to think through and across our contemporary situation, but also against the present knowledge systems that perpetuate human-centred approaches" (Oppermann 9–10). In "To Hear Plant Speak", Marder wonders about the conditions of possibility for a cross-kingdom translation and suggests that we hold a space for the untranslatable, preserving the opacity of vegetal languages:

> What kind of hearing must we resort to and what sort of speech corresponds to it? How to translate the language, or the languages, of plants into terms that are intelligible within the scope of our human languages? What is lost in this transposition, above all when it strives to make everything ... transparent? What are the conditions of possibility for a cross-kingdoms translation and what is the place of the untranslatable in it? (Marder 2017)

Marder leads us to wonder not only whether it is possible, but also whether it is desirable to translate vegetal languages. This suggests a posture of piety towards the silence of vegetal life and calls for an attunement to a speaking without words, drawing our attention to the presence of languages that we cannot ever fully comprehend nor exhaust. "Faithful to the obscurity of vegetal life, plant thinking preserves the unthinkable in its midst". Un/thinking with plants entails indeed an attention to that which escapes us—"this life's other pole, the pole of darkness with the possibilities proper to it" (Marder 2013, 173, 177). This echoes Marisol de la Cadena's advocacy for not-knowing as an analytic method: the anthropologist invites us to suspend our epistemological claims and let our knowledge be "exceeded" by "that which is beyond the limit of modern epistemological knowledge". Feel-thinking in the presence of others, including other-than-humans, while attending to their teachings might indeed "yield unexpected possibilities and the unexpected as a possibility" (de la Cadena 2021, 249, 257). Both Marder and de la Cadena wave toward our being overflowed by a meaning that escapes us, by the opaque knowledge stemming from others and by the possibilities of thought that open once we attend to this opacity. Darkness does not appear as something that should be overcome, but rather as that which light should preserve and disclose, a ubiquitous liminality, a perpetual openness into the abyss that suffers no obstruction. An apophatic attentiveness to more-than-human lives and their inscrutable ways of un/thinking, un/knowing and un/saying might thus guard us against unfortunate closures. Philosopher of religion Mary Jane Rubenstein approaches this apophasis through the prism of wonder. "Because it attends to the strangeness of the most familiar", she writes, "such wakeful thinking might finally endure, rather than close down, the perilous openness of wonder" (Rubenstein 2009, 19). Apophatic thought attends to the radical openness of more-than-humans, preserving the perpetual burgeoning of their unbounded becomings and attuned to their epistemological withdrawal, which is also an overflowing. Myers refers to the apophatic presence of vegetal beings as numinous, contending that

> it is time that we find ways to reach toward the unknowable, the imperceptible, the ineffable and the numinous. Not with the desire to capture some truth, or attempt render the world legible to the constraints of our colonized imaginations, but rather *to learn how to step into not knowing as an ethic and a practice*. Not knowing ahead of time that the trees have nothing to say. Not knowing what matters to a life or to this land. Not knowing is not about cultivating ignorance or indifference. Rather it is a capacious and humbling space that offers some refuge from the hubris of knowledge systems—like ecology—that are bound so tightly to colonial conquests, discursive regimes, cultural norms, and moral economies that have too long dictated what is good, valuable, and true. Asking "what do the trees know?", and "what matters to this land?" is thus to insist on pushing past assumptions about what counts as proper forms of knowing and the proper objects of knowledge in both anthropology and ecology. (Myers 2017, 3, my emphasis)

Pointing toward the boundaries of thought and language, plants, and with them other alien, more-than-human forms of life, summon an apophatic epistemology of un/knowing. Teaching us to un/think and regenerate the process of our thought, their earthly darkness invites us to reclaim an aesthetical, epistemological and political sensitivity to the presence and wisdom of that which escapes or is repressed from our attention. Vegetal lives signal a meaning that exceeds dominant regimes of knowledge and summon us to tend the openness of that which withdraws itself from our understanding. The apophatic concern for a preservation of this openness and resistance of the world to our desire to contain it in knowledge conjures up the Socratic maxim of ignorance: embracing withdrawal as a way of un/knowing, it cultivates the *perilous openness of wonder* in which philosophy finds its source. Reclaiming an intimacy with the darkness of vegetal lives is indeed an encounter with the openness of wonder, echoing the words of Alfred North Whitehead: "philosophy begins in *wonder*. And at the end, when philosophic thought has done its best, *the wonder remains*" (Whitehead 1938).

Bibliography

Abram, David. *The Spell of the Sensuous. Perception and Language in a More-Than-Human-World*, New-York: Vintage Books, 1996.

Abram, David and Merlin Sheldrake. "Magic and Ecology: Entangled Life", CRASSH Cambridge, 2 Apr. 2021. https://www.youtube.com/watch?v=aVpVc7VKOrM.

Barad, Karen. *Meeting the Universe Halfway. Quantum Physics and the Entanglement of Matter and Meaning*, Durham: Duke University Press, 2007.

Barad, Karen. "Diffractive Diffraction: Cutting Together-Apart", Parallax, 20:3, 2014, pp. 168–187.

Barad, Karen. "On Touching the Stranger Within—The Alterity that therefore I Am", the Poetry Project, 2018. https://www.poetryproject.org/library/poems-texts/on-touching-the-stranger-within-the-alterity-that-therefore-i-am.

Barad, Karen. "No Small Matter: Mushroom Clouds, Ecologies of Nothingness, and Strange Topologies of Spacetimemattering", in *Arts of Living on a Damaged Planet. Ghosts and Monsters of the Anthropocene*, edited by Anna Lowenhaupt Tsing, Heather Anne Swanson, Elaine Gan, Nils Bubandt, G103–120, Minneapolis: University of Minnesota Press, 2017.

De la Cadena, Marisol. "Not Knowing: In the Presence of …" in *Experimenting with Ethnography: A Companion to Analysis*, edited by Andrea Ballestero and Brit Ross Winthereik, Durham: Duke University Press, 2021.

Deleuze, Gilles and Félix Guattari. *A Thousand Plateaus: Capitalism and Schizophrenia*. Minneapolis: University of Minnesota Press, 1987.

Derrida, Jacques, "An Interview with Jacques Derrida on the Limits of Digestion." Interview by Daniel Birnbaum and Anders Olsson. *E-Flux Journal* 2, Jan. 2009. http://worker01.e-flux.com/pdf/article_33.pdf.

Gagliano, Monica, *Thus Spoke the Plant. A remarkable Journey of Groundbreaking Scientific Discoveries and Personal Encounters with Plants*, Berkeley: North Atlantic Books, 2018.

Heidegger, Martin, *Beiträge zur Philosophie* (*Vom Ereignis*) (1936–1938), Gesamtausgabe 65, edited by Friedrich-Wilhelm v. Hermann, Frankfurt am Main: Vittorio Klostermann, 2003.

Hustak, Carla and Natasha Myers. «Involutionary Momentum: Affective Ecologies and the Sciences of Plant/Insect Encounters», *differences: A Journal of Feminist Cultural Studies* 23 :5, 2012.

Keller, Catherine. *Face of the Deep: A Theology of Becoming*. London: Routledge, 2003.

Keller, Catherine. *Cloud of the Impossible: Negative Theology and Planetary Entanglement*, New-York: Columbia University Press, 2014.

Kimmerer, Robin Wall. *Braiding Sweetgrass*. Minneapolis, Minneapolis: Milkweed Editions, 2013.

Kohn, Eduardo. *How Forests Think: Toward an anthropology beyond the human.* 2013. Berkeley: University of California Press.

Marder, Michael. "Do plants pay attention? A possible phenomenological-empirical approach", *Progress in Biophysics and Molecular Biology,* 173, 11–23, 2022.

Marder, Michael. *Plant Thinking. A philosophy of vegetal life*, New-York: Columbia University Press, 2013.

Marder, Michael. "Vertimus: Ten Theses on Becoming Plant", *Anthropologie et Société,* 44:3, 2020, 195–206.

Marder, Michael. "To Hear Plants Speak", in *The Language of Plants: Science, Philosophy, Literature,* edited by Monica Gagliano, John C. Ryan and Patrícia Vieira, Minneapolis: University of Minnesota Press, 2017.

Merleau-Ponty, Maurice. *Phenomenology of Perception,* translated by Colin Smith, London: Routledge & Kegan Paul, 1962.

Mol, Annemarie. "I Eat an Apple: On Theorizing Subjectivities," *Subjectivity* 22, no. 1, 2008, 28–37.

Morton, Timothy. *The Ecological Thought.* Cambridge: Harvard University Press, 2010a.

Morton, Timothy. "The Dark Ecology of Elegy", in *The Oxford Handbook of the Elegy,* edited by Karen Weisman, 2010b.

Myers, Natasha. "Becoming Sensor in Sentient Worlds. A More-Than-Human History of a Black Oak Savannah", in *Between Matter and Method. Encounters in Anthropology and Art,* edited by Gretchen Bakke and Marina Peterson, London: Routledge, 2017.

Myers, Natasha. Interview by Carter Lou McElroy, "On Growing the Planthroposcene", *For the Wild podcast,* October 2020. https://forthewild.world/podcast-transcripts/dr-natasha-myers-on-growing-the-planthroposcene-204.

Myers, Natasha. "How to grow livable worlds: ten (not-so-easy) steps for life in the Planthroposcene", ABC Religion & Ethics, Australia Broadcasting Corporation, January 2021.

Nietzsche, Friedrich. *Daybreak: Thoughts on the Prejudices of Morality.* Translated by Maude-Marie Clark and Brian Leiter. Cambridge: Cambridge University Press, 1997.

Oppermann, Serpil. "The Scale of the Anthropocene", *Mosaic: An Interdisciplinary Critical Journal,* September 2018, 51: 3, 1–17.

Palmer, Helen and Vicky Hunter, "Worlding", *Almanac of New Materialism,* March 2018. https://newmaterialism.eu/almanac/w/worlding.html.

Pollan, Michael. *This is Your Mind on Plants,* New-York: Penguin Press, 2021.

Pollan, Michael. "The Intelligent Plant", *New-Yorker Magazine,* December 2013. https://www.newyorker.com/magazine/2013/12/23/the-intelligent-plant.

Pollan, Michael. "What It's Like to Trip on the Most Potent Magic Mushroom", *The Atlantic,* June 2018. https://www.theatlantic.com/science/archive/2018/06/what-its-like-to-trip-on-the-most-potent-magic-mushroom/561860/.

Pouteau, Sylvie. "Mouvement et monde des êtres ouverts. Vers une écologie de la représentation des plantes", *La Pensée écologique*, 6: 2, 2020.

Pouteau, Sylvie. "Plants as open beings. From aesthetics to plant-human ethics", in *Plant Ethics*, edited by Kallhoff Angela, Marcello Di Paola, Maria Schörgenhumer, London: Routledge, 2018.

Powers, Richard. *The Overstory*. London: Vintage, 2019.

Rubenstein Mary-Jane, *Strange Wonder: The Closure of Metaphysics and the Opening of Awe*, New-York: Columbia University Press, 2009.

Shechet, Elie. "Do Plants Have Something to Say?", *New-York Times*, August 2019 https://www.nytimes.com/2019/08/26/style/can-plants-talk.html.

Wandersee, James and Elizabeth Schussler, "Preventing Plant Blindness," *The American Biology Teacher*, 61, no. 2, 1999, 82–6.

Whitehead, Alfred North, "Nature Alive": Lecture Eight in *Modes of Thought*. New York: Macmillan, 1938, 202–232.

CHAPTER 10

To Be a Pixel Flower: Player-Plant Entanglement in Video Games

Carolin Becklas

1 The Profound Responsibilities of Plants

Without flowering plants, there would be no us. Their revolution started in the Cretaceous, in the last 70 million years of the age of dinosaurs and it dramatically boosted Earth's biodiversity on land.[1] More than a million species of modern insects such as bees, butterflies, bugs, beetles, wasps, and flies, owe their livelihoods to flowering plants. Plants also sustain us humans: We eat them, we wear them, we use them for medicine, energy, cosmetics, and building materials. In short, without them, we would not be able to survive. We share a multifaceted relationship extending far back into our joint evolutionary history[2] but somehow, we have not yet learned how to live together sustainably.

The utter indispensability of plants for our entangled lives on this planet is still difficult for us to put into words or to, for example, simulate in vegetation models used for climate simulations. Emily Dickinson, however, who spent many hours in her garden observing exactly which flowers thrive where and under what conditions,[3] was able to express these intricate entanglements seemingly effortlessly in just two verses: "To be a Flower, is profound/ Responsibility—."[4]

Despite our entangled lives, we rarely think about plants—and if we do, we tend to think of them as commodities, landscapes, metaphors, or food. Plants do not display obvious emotional or behavioral responses to their

1 Michael J. Benton, Peter Wilf, and Hervé Sauquet, "The Angiosperm Terrestrial Revolution and the Origins of Modern Biodiversity," *The New phytologist* 233, no. 5 (2022), https://doi.org/10.1111/nph.17822.

2 Barbara Schaal, "Plants and People: Our Shared History and Future," *PLANTS, PEOPLE, PLANET* 1, no. 1 (2019), https://doi.org/10.1002/ppp3.12.

3 Judith Farr and Louise Carter, *The Gardens of Emily Dickinson* (Cambridge, Mass.: Harvard University Press, 2004).

4 Emily Dickinson, "Bloom," in *The Complete Poems*, ed. Emily Dickinson and Thomas H. Johnson, [Repr.], Back Bay Books (Boston, Toronto, London: Little, Brown and Company, 1960) lines 15–16.

 | DOI:10.1163/9789004721777_012

environment, which can make it more challenging for us to empathize with them. Additionally, humans have historically viewed plants as passive, inanimate objects that exist primarily for human use and consumption. Aristotle's characterization of plants has been particularly influential in informing the manner people from the Global North imagine and represent plants in art and literature.[5] While he recognizes that plants are dependent on sunlight, air, and water for their survival, and that they have mechanisms for responding to changes in their environment, such as by growing towards sources of light or moisture, he classifies them as intermediaries on a spectrum of sentience. This perspective has led to a scarcity of plant-subjects in Western literature and art and to a lack of understanding and appreciation for the complex biological and ecological systems that plants are a part of, and the ways in which they contribute to the health and well-being of our planet. Simultaneously, however, it may have also charged "the vegetable kingdom with an uncanny ontological potency."[6] Plants have an uncanny quality to them precisely because we (Europeans and North Americans) struggle to imagine plants as cognitive, learning, procreating, communicating subjects. The French botanist and plant biologist Francis Hallé therefore argues that "plants represent absolute otherness to us."[7]

In a time of global warming, however, our vegetal neighbors are becoming increasingly hard to ignore. They do not only "dominate every terrestrial environment, composing ninety-nine percent of the biomass on Earth"[8] but they also play critical roles in mitigating climate change and promoting the overall well-being of our planet.[9] It seems, as our understanding of our own vulnerability has grown, so has our interest in plants. This shift in perspective provides the breeding ground for a new research direction in biology that, since the founding of the *Society for Plant Neurobiology* in 2005, has challenged some fundamental assumptions about plants. This relatively new field of inquiry called plant neurobiology, addresses the question of the extent to which plants are intelligent—although, as Trewavas has rightfully pointed out, intelligence is a fraught term due to its anthropocentric slant placed on the meaning of the

5 Aristotle, R. Cresswell and J.G. Schneider, *Aristotle's History of Animals: In Ten Books* (London: George Bell and Sons, 1901).

6 Randy Laist, "Introduction," in *Plants and Literature: Essays in Critical Plant Studies*, ed. Randy Laist, Critical plant studies: philosophy, literature, culture [1] (New York, NY: Editions Rodopi, 2013), 12.

7 Francis Hallé, *In Praise of Plants* (Portland, Or.: Timber Press, 1999), 37.

8 Michael Pollan, "The Intelligent Plant," *The New Yorker*, December 16, 2013, accessed April 8, 2023, https://www.newyorker.com/magazine/2013/12/23/the-intelligent-plant.

9 Giuseppe Di Vita et al., "A Review of the Role of Vegetal Ecosystems in CO2 Capture," *Sustainability* 9, no. 10 (2017), https://doi.org/10.3390/su9101840.

word.[10] Plant neurobiology researchers are concerned with various aspects of plant behavior, such as how plants sense and respond to light, gravity, touch, and other stimuli, as well as how they communicate with other plants and organisms in their environment.

Critical plant studies, a "broad framework for re-evaluating plants, their representations, and human-plant interactions,"[11] benefits from these new insights into the lives of plants. While the various representations and symbolic functions of plants are now a vibrant research topic in literary studies, researchers from game studies still seem to suffer from "plant blindness."[12] This is quite surprising as plants are frequently points of discussion in all areas orbiting the billion dollar gaming industry: Players, for example, appreciate realistic in-game vegetation, even eagerly discuss the best-looking grass in video games,[13] which is why realistic physical rendering of gameworld elements has become a major selling point for newer video games,[14] prompting game studios to hire their own vegetation artists whose job it is to create (realistic) plant life assets for in-game environments only. And, finally, plants fulfill important narrative, visual, and gameplay functions in the games themselves, which will be the focus of this paper.

Unlike physical books, which are quite literally dependent on plants because they are printed on paper, this material connection is cut in digital representation. Lykke Guanio-Uluru, who researches plant representation in children's literature, calls attention to this breach and asks: "Do plants become *more* visible when they are *not* the medium of representation, or less?"[15] Instead of providing a definite yes or no answer to Guanio-Uluru's question, this paper,

10 Anthony Trewavas, "Aspects of Plant Intelligence," *Annals of botany* 92, no. 1 (2003), https://doi.org/10.1093/AOB/MCG101.

11 Anna M. Lawrence, "Listening to Plants: Conversations Between Critical Plant Studies and Vegetal Geography," *Progress in Human Geography* 46, no. 2 (2022): 630, https://doi.org/10.1177/03091325211062167.

12 James H. Wandersee and Elisabeth E. Schussler, "Preventing Plant Blindness," *The American Biology Teacher* 61, no. 2 (1999), https://doi.org/10.2307/4450624.

13 See, for example, Eric Schulkin, "He Says It's the 'Best Grass' in Video Games, Daring Other Devs to One-up Him." *Kotaku*, March 18, 2022. Accessed April 8, 2023. https://kotaku.com/lushfoil-photography-simulator-steam-pc-tiktok-1848666248; Kevin Foley, "Here's Some of the Best Grass in PC Gaming (And Where to Get It)." *PC Invasion*, April 20, 2021. Accessed April 8, 2023. https://www.pcinvasion.com/best-grass-pc-gaming-graphics/; Stylized Station, *Top 10 Grass in Video Games (Yes I'm Serious)*., 2022. YouTube video. Accessed August 25, 2023. https://www.youtube.com/watch?v=FDV7Jz8LexM&t=3s.

14 Scott Roberts and Dale Patterson, "Virtual Weather Systems," in *Proceedings of the Australasian Computer Science Week Multiconference* (New York, NY, USA: ACM, 2017).

15 Lykke Guanio-Uluru, "Analysing Plant Representation in Children's Literature: The Phyto-Analysis Map," *Children's Literature in Education*, 2021, https://doi.org/10.1007/s10583-021-09469-2 under "Petal 1, Mediation," italics in original.

following Dickinson, will rather operate with a flexible spectrum of varying degrees of "responsibilization" of plants in digital games. My phytocentric analysis will foreground plants as the objects of analysis and ask how the often unimaginable live of plants is imagined and "played" in a few selected examples. How are plants visually represented and mechanically simulated? What is their (narrative) function and status in the games? Are they imagined as agents in their own right? This paper is therefore concerned with cataloging, analyzing, and discussing the varying "responsibilities," representations, and functions plants assume in games.

2 Pixelated Worlds and the Crisis of Imagination

In his seminal study of the American nature writing tradition, Lawrence Buell postulates that the "environmental crisis involves a crisis of the imagination the amelioration of which depends on finding better ways of imaging nature and humanity's relation to it."[16] Buell is arguing that solving the environmental crisis requires not just technological fixes, but also a shift in our cultural imagination to see ourselves as part of, rather than separate from, nature. This shift in imagination would require us to rethink our values, beliefs, and practices in relation to the environment and to develop new ways of imagining our place in the natural world that are not based in appropriative and hegemonic relations to the more-than-human realm. This also includes recognizing our entangled fates with plants on an increasingly warming planet.

Digital games, defined as imaginative, interactive, and exploratory spaces, shape and are shaped by these environmental discourses.[17] Essentially, all games are environments,[18] and these digital environments make claims about "how something does, should, or could work"[19] and are steeped in operations

16 Lawrence Buell, *The Environmental Imagination: Thoreau, Nature Writing and the Formation of American Culture*, 2. printing (Cambridge, Mass., London: Harvard University Press, 1995), 2.

17 Hans-Joachim Backe, "Greenshifting Game Studies: Arguments for an Ecocritical Approach to Digital Games," accessed September 8, 2022, http://www.firstpersonscholar.com/greenshifting-game-studies/.

18 Alenda Y. Chang, *Playing Nature: Ecology in Video Games*, Electronic mediations 58 (Minneapolis: University of Minnesota Press, 2019).

19 Ian Bogost, *Persuasive Games: The Expressive Power of Videogames* (Cambridge, Mass., London: MIT Press, 2010), 58.

of power. Moreover, as games are highly multimodal artifacts,[20] game developers can compose gameworlds using virtually all semiotic modes, which can be defined as "a socially shaped and culturally given resource for making meaning."[21] Representations of plants can therefore be constructed multimodally, including but not limited to moving and still image, color, sound effects, music, text, layout, and haptics.

Unlike in literature, movement is often an essential part of experiencing the fictional gameworld. Navigating through space is what enables players to influence the virtual environment, which is what allows the narrative to unfold: "[T]he story is tied to the navigation of space."[22] In gameworlds, the player updates in real time some of the affordances offered by the software and the rules inscribed within it, and thus generates their own version of the world during each playthrough. The interactive nature of these worlds sets them apart from the fictional worlds of films, television, and novels. These latter worlds are self-contained and may be "read" multiple times, while gameworlds are in *statu nascendi*, as each playthrough is similar but not identical. Players appropriate these worlds with the help of their avatars. Their representatives in the game allow the players to participate in it, an important difference to film, for example:

> If a film audience were to step through the camera and onto the film set they would see a modern film studio [...] This space is not the world of the story but that of the production of the film. The illusion created by the fictional world would be broken. In contrast, it is a defining characteristic of video game spaces that they allow this step into the represented space. The result is a hybrid between architectural navigable and cinematically represented space.[23]

However, these gameworlds are anything but "ready-made." Rather, they are always already woven into a whole range of parallel, culturally shaped world versions. We use our imagination and knowledge about already existing worlds

20 Jason Hawreliak, *Multimodal Semiotics and Rhetoric in Videogames*, Routledge studies in multimodality (New York, London: Routledge, 2019).

21 Gunther Kress, *Multimodality: A Social Semiotic Approach to Contemporary Communication* (New York: Routledge, 2010), 79.

22 Janet Murray, *Hamlet on the Holodeck: The Future of Narrative in Cyberspace* (New York: Free Press, 1997), 132.

23 Michael Nitsche, *Video Game Spaces: Image, Play, and Structure in 3D Game Worlds*, The MIT Press (Cambridge, Mass.: MIT Press, 2008), 85.

to decompose "old" worlds and to compose new ones[24]—this applies to the fictional worlds of film, literature, and art as well as to the pixel worlds of video games. The imaginaries of the more-than-human world is therefore confined to the limits of the human imagination, which includes our languages, images, sounds, and so on. There are, however, profound differences between languages and their capabilities to imagine, frame, and represent the interconnectedness between the human and more-than-human world as the Native American botanist and author Robin Wall Kimmerer has described in her book "Braiding Sweetgrass: Indigenous Wisdom, Scientific Knowledge and the Teachings of Plants."[25] According to Kimmerer, indigenous languages have more complex ways of describing the natural world than European languages and often use a relational grammar, which emphasizes the relationships between things rather than their individual qualities. And, perhaps most importantly, "we use the same words to address the living world as we use for our family. Because they are our family."[26] Linguistically including plants in one's family also means perceiving them as people rather than objects. Some European languages, however, refer to a tree as an "it", and therefore, "we make that maple an object; we put a barrier between us, absolving ourselves of moral responsibility and opening the door to exploitation. Saying it makes a living land into 'natural resources.' If a maple is an it, we can take up the chain saw. If a maple is a her, we think twice."[27]

3 Plants as Resources

In games, prosperity and happiness often start with a chain saw—or its little brother, the hatchet. In the highly popular sandbox game "Minecraft"[28] in which players explore a 3D blocky world and have the ability to mine and craft various types of blocks to build structures, tools, and other items, one of the first in-game actions is to chop down trees to collect wood, which is essential for building tools and shelter. Similarly, the "Civilization" series[29] begins with the seemingly simple act of chopping down a tree, which is used to build a farm

24 Nelson Goodman, *Ways of Worldmaking* (Indianapolis: Hackett, 1978).

25 Robin Wall Kimmerer, *Braiding Sweetgrass: Indigenous Wisdom, Scientific Knowledge and the Teachings of Plants* (Minneapolis, Minnesota: Milkweed Editions, 2013), https://ebookcentral.proquest.com/lib/kxp/detail.action?docID=1212658.

26 Ibid., 55.

27 Ibid., 57.

28 Mojang Studios, *Minecraft* (Mojang Studios; Sony Computer Entertainment, 2011).

29 Sid Meier and MicroProse, *Civilization* (MicroProse, 1991).

that grows into a town. As the town grows, more trees are chopped down, more farms and towns are built, and resources like stone and metal are extracted to create new technologies and raise armies. The player's goal is to capture more cities, and in doing so, they raze more forests, build more farms, and dig more mines to satisfy the needs of the expanding empire. These digital, arboreal imaginaries therefore highlight the destruction of trees and forests in the name of human industry and expansion, in which the "demand for timber is both the condition for and effect of an imperial desire."[30] This growth imperative is often encoded through game mechanics and win conditions in game genres that focus on building, managing, and expanding a city, an empire or civilization, such as sandbox games, simulation games, or strategy games. In these cases, winning the game often includes constant resource acquisition, which is therefore rewarded and encouraged through the game's mechanics. Ludic player-tree entanglement is especially crucial for success, as trees provide the building and crafting materials for almost all in-game structures. Therefore, trees are the skeleton of playful interaction upon which everything else hangs.

In other genres, such as adventure or role-playing games, harvesting plants is vital for the avatar's survival as they provide food, medicine, and materials to craft and upgrade the inventory and the player's stats. Thus, the necessary mechanics are often explained to the players right at the beginning of the game. One prominent example is the "lessons of the wild" which are taught to the protagonist Aloy in "Horizon Zero Dawn,"[31] an action role-playing game that boasts a global sales figure surpassing 10 million copies by 2021. Within these instructional sessions, the game's main character Aloy is taught crucial survival techniques and wisdom by her adoptive father, Rost. Some of his lessons include how to gather resources and make use of them, and how to find and use natural cover in tall grass. Aloy is educated about the vital significance of medicinal berries, flowers, and plants, which can serve as both life-sustaining and life-saving assets, which is why Rost instructs her to always have her pouch full of them. Without these vegetal resources, the player cannot win the game.

Thus, much like in the "real" world, plants in video games are givers of life, food, shelter, and fire. When plants are treated as infinite resources in games, they are tools for the avatar's survival and the player's fantasies of mastery and expansion. Across genres, the game mechanics often encourage the endless

30 Solvejg Nitzke and Helga G. Braunbeck, "Arboreal Imaginaries. An Introduction to the Shared Cultures of Trees and Humans," *Green Letters* 25, no. 4 (2021): 342, https://doi.org/10.1080/14688417.2021.2072633.

31 Guerrilla Games, *Horizon Zero Dawn* (Sony Interactive Entertainment, 2017).

accumulation of resources and construct nature as an inexhaustible repository for the player to take advantage of. While plants shoulder a high degree of responsibility for success, their status and agency in the gameworld is marginal. They provide the inanimate, constantly respawning (i.e., reappearing) material necessary for progression in the game but do not exert any force on their surroundings.

4 Plants as Backdrop and (Aesthetic) Pleasure

Unsurprisingly, most ludic representation of plants is anthropocentric and figures them as inanimate part of the digital landscape, the "gamescape,"[32] or as the backdrop for the in-game action and narrative. They seldom encourage curiosity about plants and obscure the artistry that goes into designing complex digital flora within a limited computer memory budget.

During the COVID-19 pandemic, gardening games, which offer a sense of calm and relaxation, became particularly popular as people sought out games that could provide an escape from the stress and uncertainty of the world around them.[33] Especially "Animal Crossing: New Horizons,"[34] whose release coincided with the onset of several lockdowns across the globe, benefited from the exceptional situation and became the second-best selling video game ever in Japan.[35] In the game, players take on the role of a customizable human character who moves to a deserted island to start a new life. The main gameplay mechanic revolves around catching wildlife, donating fossils and species to a museum and growing vegetation. While plants have some functional purpose and require some level of care and attention from the player, they mostly play a cosmetic role as they enhance the overall aesthetic of the player's island. Plants therefore oscillate between functional design object on the one hand and work of art on the other, creating some blissful moments of aimless strolling in lush pixel gardens.

---

32 Shoshana Magnet, "Playing at Colonization," *Journal of Communication Inquiry* 30, no. 2 (2006): 142.

33 M.Á. López-Cabarcos, Domingo Ribeiro-Soriano, and Juan Piñeiro-Chousa, "All That Glitters Is Not Gold. The Rise of Gaming in the COVID-19 Pandemic," *Journal of Innovation & Knowledge* 5, no. 4 (2020), https://doi.org/10.1016/j.jik.2020.10.004.

34 Nintendo, *Animal Crossing: New Horizons* (Nintendo, 2020).

35 L. Doolan, "Animal Crossing: New Horizons Is Now Japan's Second Best-Selling Game Ever," Accessed April 8, 2023, https://www.nintendolife.com/news/2020/08/animal_crossing_new_horizons_is_now_japans_second_best-selling_game_ever.

FIGURE 10.1 Plant expressing satisfaction in "Luxuria Superbia"

Curiously absent in these gardening games are death, decay, and decomposition which are a natural part of any non-digital garden. However, as the games usually focus on aesthetic pleasure and providing comfort and escape, the natural processes and multimodal representation of the plant's life cycle is kept to a minimum.

Instead of providing pleasure for the human eye, the game "Luxuria Superbia"[36] lets players pleasure plants, thus turning power relations and scopic regimes upside down. With the opening screen, the game instructs the player: "Touch me" is written in bright letters—and this is meant to be understood literally. The game is played by touching and stroking the screen of a device, and the player interacts with an abstract flower-like structure that grows and changes as they progress through the levels. The designers took full advantage of the many modes available to them in digital games. Vibrant colors light up and the soundtrack intensifies as the player pleasures the plant, which also starts speaking to the player suggestively (see fig. 10.1). After the completion of a level, the flower rates the player's performance. The higher the time invested and the more skillful the use of (multiple!) fingers, the happier the plant. The game's title, "Luxuria Superbia," translates to "Prideful Lust" in Latin, which reinforces the sexual connotations of the abstract, colorful, and flowing visuals that can be interpreted as suggestive of sexual organs, particularly female genitalia. Reaching climax results in the collection of seeds for a rather unusual garden with a temple that starts out blank and becomes more colorful once the

36 Tale of Tales, *Luxuria Superbia* (Tale of Tales, 2013).

player has successfully brought pleasure and joy to the various flowers of their garden. While "Luxuria Superbia" is a commendable example of "rambunctious," clever gameplay,[37] mainstream gaming demonstrably relegates plants to the background and often incorporates them as props rather than agents.

5 Plants as Companions and Antagonists

For thousands of years, trees have not only provided us with shelter, protection, and food, but also with an abundance of resources for medicine, fire, energy, weaponry, toolmaking, and construction. They are prominently featured in mythology and folklore, where they are often represented as powerful symbols of connection, growth, regeneration, and knowledge, embodying the cyclical nature of life and the interconnectedness of all living things. In many cultures, trees are also associated with wisdom and knowledge. The Norse god Odin is said to have gained wisdom by hanging himself from the world tree Yggdrasil for nine days, while the Buddha is said to have achieved enlightenment under the bodhi tree.

Throughout the Northern Hemisphere, oaks in particular have been appropriated in symbolic representations,[38] where they have been associated with traits such as longevity, strength, fertility, and robustness. They hold important cultural, symbolic, and emotional value in many European cultures and are linked with historical stories and myths. The "German forest," and the "German oak", for example, were once central components of the political imaginary and are still present in cultural memory, yet repeatedly overwritten by varying connotations that accompany changes in collective conceptions of identity.[39]

Ancient, mystical, and wise oaks have also taken strong roots in the digital realm of video games. Since plant organisms reach an age of several hundred years, they are ahead of humans in experience and knowledge. In the realm of the imaginary, they can communicate this wealth of experience linguistically,

37 Alenda Y. Chang, "Rambunctious Games: A Manifesto for Environmental Game Design," *Art Journal* 79, no. 2 (2020), https://doi.org/10.1080/00043249.2020.1765557.

38 Thibault Leroy, Christophe Plomion, and Antoine Kremer, "Oak Symbolism in the Light of Genomics," *The New phytologist* 226, no. 4 (2020), https://doi.org/10.1111/nph.15987.

39 Evi Zemanek, "Pflanzliche Sinnbilder. Satirische Kritik Am Deutschen Wald Und Der Deutschen Eiche Als Symbole Für Nationale Identität Und Kulturelle Nachhaltigkeit," in *Literaturen Und Kulturen Des Vegetabilen. Plant Studies—Kulturwissenschaftliche Pflanzenforschung*, ed. Urte Stobbe, Anke Kramer and Berbeli Wanning, Studies in Literature, Culture, and the Environment v. 10 (Frankfurt a.M.: Peter Lang GmbH, Internationaler Verlag der Wissenschaften, 2022).

FIGURE 10.2 The Deku Tree in two of its manifestations. Left: "Ocarina of Time," right: "Breath of the Wild"

as in the "Legend of Zelda" series,[40] which features one of the most prominent examples of a talking oak. The so-called Deku Tree appears in multiple games and is a sentient being that offers guidance and support to the protagonist, Link, throughout his journey. It is depicted as a fatherly, nurturing figure which provides protection to the forest and its inhabitants. The Deku Tree is grand and imposing, with its thick trunk and leafy canopy creating an atmosphere of awe and reverence. Eyes, eyebrows, nose, and lips are clearly distinguished by bulging protrusions in the digital bark (see fig. 10.2).

In "The Legend of Zelda: Breath of the Wild,"[41] the tree explains to Link what happened 100 years ago when the war against evil was lost and what task will be bestowed upon him to restore peace. It remains unclear how many years the tree has been alive, but its wealth of wisdom spans centuries, from which the player now also benefits, as it provides them with central elements of the game's narrative and various items essential to complete the game.

However, trees and other plants may also serve as antagonists, attacking the player and thereby hindering their progression in the game. The most iconic "killer plant" is probably the so-called "Pakkun Flower" from Nintendo's "Super Mario" franchise media, which is commonly depicted as a leafy green stalk topped with a white-spotted red or green globe. They are most often attached to green pipes scattered throughout the gameworld and their mouth is lined with sharp teeth that resemble those of piranhas, thereby visually signaling danger to the player. If the player fails to defeat the plant by jumping on it or throwing objects at it, Mario takes damage and loses a portion of his health.

---

40 Nintendo, *The Legend of Zelda* (Nintendo, 1986).

41 Nintendo, *The Legend of Zelda: Breath of the Wild* (Nintendo, 2017).

In this case, the plant is an annoyance rather than an important element of the game's visual, narrative, or procedural design.

There are, however, also examples where "killer vegetables" possess much more multimodal potency. In Naughty Dog's highly popular action-adventure "The Last of Us,"[42] for example, the player is plunged into a dystopian fungal future, in which the Cordyceps fungus has mutated and spread to humans, causing widespread death, destruction, and chaos.[43] The game takes inspiration from the real-life Cordyceps fungus, a genus of fungi that infects insects and other arthropods. It is also known as the "zombie fungus" due to its ability to take over the host's body and manipulate its behavior for the purpose of spore dispersal. The infected host is often killed in the process. The game plays with the idea of a man-eating fungus and anthropomorphizes the fungal infection in the most literal sense: The digital version of the Cordyceps fungus infects human hosts and turns them into so-called "clickers," aggressive and dangerous human-fungi hybrids that relentlessly pursue non-infected individuals. Besides hijacking the host's brain and altering their behavior, the infection also causes the host's body to transform. Once infected, the body is depicted as being covered in fungal growths, with the fungus visibly penetrating the flesh and altering the appearance of the infected. As soon as a clicker completes its life cycle, spores effuse from the dead host's body and infect the surroundings, thus turning the environment into a potential source of infection. It becomes a lethal trap for flora and fauna, infecting and killing plants and animals and altering the gamespace.

Unlike some environmental writing that depicts humanity's abuse of natural resources as leading to a reckoning, however, the game does not present the Cordyceps infection as a punishment.[44] It is simply a power-play between fungi and humans that is won by kingdom Fungi. During the game, humans are powerless to control the fungus, find a cure, or prevent its spread and are left with nothing more than the hope that over time, humans will be born with immunity to the infection. The game envisions a radical player-fungi entanglement, one that is unescapable, leading to a future in which humans are no longer on the top of the Great Chain of Being.

Such "killer plant" narratives can be found in many cultures and are a common theme in movies and literature, where they make for intriguing,

42 Naughty Dog, *The Last of Us* (Sony Computer Entertainment, 2013).

43 Although fungi are their own taxonomic classification, they share obvious similarities with plants, which is why I decided to include them in this paper. Both fungi and plants are immobile, have rigid cell walls, and grow in soil.

44 Amy M. Green, "The Reconstruction of Morality and the Evolution of Naturalism in the Last of Us," *Games and Culture* 11, 7–8 (2016), https://doi.org/10.1177/1555412015579489.

sometimes horrifying, storylines in which the line between predators and prey becomes blurry. Carnivorous, human-eating plants in video games make us shiver because they challenge our understanding of the man-nature hierarchy and reverse established human-plant orders. If plants suddenly seek revenge for what we have done to them, how safe can we sleep in our (often wooden) beds?

6 Plants as Evidence for and Mitigation of Digital Global Warming

While we humans can only experience the weather that currently prevails, trees are living archives of past climates.[45] By examining the rings of a single tree, it is possible to identify environmental stressors that the tree experienced during its lifetime, such as drought or fire. Additionally, tree rings usually grow wider in warm, wet years and are thinner in cold and dry ones, making them sensitive proxies of past climate change.

In games, plants also indicate the environmental status of the gameworld and even act as the "lungs of the virtual Earth" by exchanging oxygen and carbon dioxide with the atmosphere. In the game "Eco"[46] by Strange Loop Games, for example, trees play a central role in the player's interactions with the environment. In this multiplayer online game, players must collaborate to build a civilization and prevent ecological disaster. Players must balance the needs of their growing population with the impact of their industrial and agricultural activities on the environment, as well as deal with the consequences of climate change and other natural disasters. Trees are an important in-game resource, but they also regulate the game's intricate, and highly responsive ecosystem. "Eco" utilizes an elaborate climate model and not only simulates a planet with different biomes and climate zones, but also a pollution system that consists of air and ground pollution components. Prolonged air and ground pollution leads to sea level rising and an increase in baseline temperature which has devastating effects on the environment and eventually on the virtual community who need food and drinking water. The careful growth of trees can help to absorb carbon dioxide from the atmosphere, which can help to mitigate the effects of digital global warming. On the other hand, clear-cutting forests can lead to increased greenhouse gas emissions, which can contribute to climate change and other environmental risks. A balanced approach to forest

45 Virginia Gewin, "The Trees' Lessons: Climate Records Are Written in Tree Rings," *Nature* 612, no. 7938 (2022), https://doi.org/10.1038/d41586-022-04150-0.

46 Strange Loop Games, *Eco* (Strange Loop Games, 2018).

management that considers the needs of the community as well as those of the environment is therefore key here.

Thus, "Eco" offers a virtual playground for developing and applying various strategies and solutions to environmental problems over which players often have little or no decision-making power in everyday life. In games like "Eco," which focus on the playful communication of climate science and sustainability issues, plants play a vital role in regulating harmful gases and are thus an integral part of the digital Earth's well-being. In these cases, pixel plants perform similar, albeit less complex, tasks as their real-world siblings. Thus, although plant characters are often not playable in these cases, to be a pixel plant in "sustainability games" is indeed profound responsibility.

7 Plants as Procedural Practice and Design

The arboreal roots have, however, not only migrated into the design and story of various playable and non-playable characters in video games but have also burrowed into the structure of the games themselves. As Andrew Piper writes, "[i]nside of every computer, at the very bottom of it all, lies a tree."[47] This tree structure is a foundational organizing principle for the way computers operate, which consists of a hierarchy of folders and files, much like the branches and leaves of a tree. Every computer uses this tree-like structure to organize, store and access data in a systematic manner, be it "the bifurcating nature of the if/then statement or the computational logic of the binary tree."[48]

It is, of course, wrong to assume that the organization of human knowledge into tree-like structures is an innovation of the computer age. Although trees became popular as a method of communicating knowledge with Charles Darwin, "tree diagrams were used by a variety of philosophical, religious, and secular scholars to depict phenomena such as 'logical relationships,' 'affiliations,' 'genealogical descent,' 'affinity,' and 'historical relatedness' between the elements portrayed on the tree,"[49] even before Darwin. Its iconography crystallizes basic human desires for order, stability, and hierarchy.

---

47 Andrew Piper, *Book Was There: Reading in Electronic Times* (Chicago, London: University of Chicago Press, 2013), 125.

48 Ibid.

49 Nathalie Gontier, "Depicting the Tree of Life: The Philosophical and Historical Roots of Evolutionary Tree Diagrams," *Evolution: Education and Outreach* 4, no. 3 (2011): 515, https://doi.org/10.1007/s12052-011-0355-0.

Particularly in the case of strategy or so-called 4X-games, i.e., games whose mechanics encourage exploration, expansion, exploitation, and extermination, tree-like graph structures are used to visualize the hierarchical structures of development possibilities within the gameworld. These visualizations are commonly referred to as technology trees, or tech trees, and have been used in the gaming industry for three decades.[50] They can loosely be defined as "a structure that controls progress from one technology to a better technology, enabling the player to create better facilities or more powerful units."[51] In a tech tree, each technology or upgrade is represented as a node on a branching diagram (see fig. 10.3). Players must research each technology in order, unlocking access to more advanced options as they progress through the tree. Typically, each technology or upgrade requires a certain number of resources or time to research, creating a strategic trade-off for the player between investing in new technologies and using those resources to build up their current capabilities.

In Sid Meier's "Civilization" series, the player's task is to build the most "successful" civilization possible from the prehistoric past of Neolithic times through human history to the present and future colonization of the universe. Prominently displayed in the interface design of each game is a highly elaborate tech tree that players must carefully navigate to progress through different eras and achieve victory. The tech tree in "Civilization IV,"[52] for example, is divided into four eras: Ancient, Medieval, Industrial, and Modern, with each era containing a set of technologies that the player must research to unlock the next era's technologies. The game's procedural tree-like structure therefore dictates the chronological development and progression through the game. Following the procedural branches of the tree ultimately ends with democracy and capitalism, hence presenting the player with a Euro-centric master narrative that guarantees the success of Western-oriented culture.[53] As Gontier suggests through her analysis of the historical roots and cultural meanings of tree diagrams, these hierarchical representations were the outgrowth of ancient philosophical attempts to find the "true order" of the world and to map

---

50 Teemu J. Heinimäki, "Technology Trees in Digital Gaming," in *Proceeding of the 16th International Academic MindTrek Conference*, ed. Artur Lugmayr (New York, NY, USA: ACM, 2012).

51 Dave Morris and Leo Hartas, *Strategy Games*, Game guru 1 (Lewes, England: Ilex, 2004), 141.

52 Sid Meier and Firaxis Games, *Civilization IV* (2K, 2005).

53 Kacper Pobłocki, "Becoming-State. Bio-Cultural Imperialism of Sid Meier's Civilization," *Focaal—European Journal of Anthropology* 39 (2002).

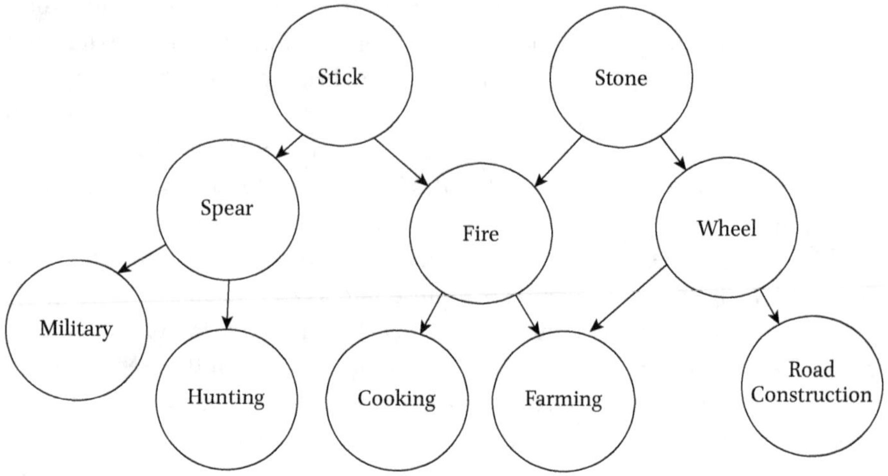

FIGURE 10.3 Example of a simple tech tree

the world "as it is."[54] Games like "Civilization IV" feed into this narrative of ultimate order by depicting the world as inherently manageable, provided the right technological "advancements" are unlocked, thus appealing to our bedrock faith in technology. Our actions and thinking in the Anthropocene tend to over-rely on technological fixes that produce unintended consequences. While these may be appealing at first glance due to their ability to provide straightforward and effective solutions to complex environmental problems, they do not address the underlying political, economic, or social causes of such problems.[55]

Tech trees in games are therefore always interwoven with questions of power and influence while also, as a procedural choice, guaranteeing certain rewards for the player, thus encouraging them to continue playing—an important feature for game designers and publishers alike.

8 Conclusion

Since the beginning of the gaming industry, video games have been teeming with flora and fungi but only a handful have built meaningful gameplay mechanics around them. Rather, plants in games are mostly generic healing

54 Gontier, "Depicting the Tree of Life: The Philosophical and Historical Roots of Evolutionary Tree Diagrams."

55 Howard P. Segal, "Practical Utopias: America as Techno-Fix Nation," *Utopian Studies* 28, no. 2 (2017), https://doi.org/10.5325/utopianstudies.28.2.0231.

items or non-interactive elements of the gameworld. It seems that in a medium so dependent on movement, translating the sessile lifestyle of plants into a playable experience comes with a few challenges. This might also be the reason why animal protagonists have already made the leap into the digital world several times, despite their differences from humans. For plant protagonists, however, it might be difficult to follow in the footsteps of their animal counterparts due to their "otherness" in terms of sensory perception, mobility, or nutrition. Creating meaningful plant agency within digital ecosystems is thus always a thought experiment as it radically challenges these very notions about mobility, reproduction, sentience, and representation and thereby disrupts anthropocentric concepts of human-nature relations. It seems that this thought experiment is still too risky for many in the industry. As a result, gameworlds, unfortunately, have so far only very rarely been populated by significant vegetal characters with unique abilities, needs, and desires. Likewise, games seldom incorporate gameplay mechanics that encourage players to consider the impact of their actions on the environment, such as resource management and sustainability.

As mentioned before, there has been little cross-pollination between critical plant studies and game studies to date. Above all, there is a lack of reliable descriptive tools and methodological systems that do justice to the specificity of the subject matter and to the medium's own distinct affordances. However, the indifference of game developers and game scholars to plants in digital games is becoming increasingly untenable in the face of climate change and the omnipresence of games in our private and public spaces. Due to games unique affordances, they cannot only challenge players to entertain alternative human-plant relationships through play but may also provide them with the tools to enact alternative ontologies and meaningful player-plant entanglement. This does not suggest that incorporating more meaningful plant action in games provides easy solutions to our many environmental problems, which we have brought upon ourselves, but that they more radically start challenging our anthropocentric view of our role in an interwoven ecosystem or, at the very least, stop being a big part of the problem. Re-thinking human-plant entanglement in the digital realm does not only mean appreciating the manyfold responsibilities of plants but also accepting our own profound responsibility on an increasingly hot planet.

Bibliography

Aristotle, R. Cresswell, and J.G. Schneider. *Aristotle's History of Animals: In Ten Books*. London: George Bell and Sons, 1901.

Backe, Hans-Joachim. "Greenshifting Game Studies: Arguments for an Ecocritical Approach to Digital Games." Accessed September 8, 2022. http://www.firstpersonscholar.com/greenshifting-game-studies/.

Benton, Michael J., Peter Wilf, and Hervé Sauquet. "The Angiosperm Terrestrial Revolution and the Origins of Modern Biodiversity." *The New phytologist* 233, no. 5 (2022): 2017–35. https://doi.org/10.1111/nph.17822.

Bogost, Ian. *Persuasive Games: The Expressive Power of Videogames*. Cambridge, Mass., London: MIT Press, 2010.

Buell, Lawrence. *The Environmental Imagination: Thoreau, Nature Writing and the Formation of American Culture*. 2. printing. Cambridge, Mass., London: Harvard University Press, 1995.

Chang, Alenda Y. *Playing Nature: Ecology in Video Games*. Electronic mediations 58. Minneapolis: University of Minnesota Press, 2019.

Chang, Alenda Y. "Rambunctious Games: A Manifesto for Environmental Game Design." *Art Journal* 79, no. 2 (2020): 68–75. https://doi.org/10.1080/00043249.2020.1765557.

Di Vita, Giuseppe, Manuela Pilato, Biagio Pecorino, Filippo Brun, and Mario D'Amico. "A Review of the Role of Vegetal Ecosystems in CO2 Capture." *Sustainability* 9, no. 10 (2017): 1840. https://doi.org/10.3390/su9101840.

Dickinson, Emily. "Bloom." In *The Complete Poems*. Edited by Emily Dickinson and Thomas H. Johnson. [Repr.], 482. Back Bay Books. Boston, Toronto, London: Little, Brown and Company, 1960.

Doolan, L. "Animal Crossing: New Horizons Is Now Japan's Second Best-Selling Game Ever." Accessed April 8, 2023. https://www.nintendolife.com/news/2020/08/animal_crossing_new_horizons_is_now_japans_second_best-selling_game_ever.

Farr, Judith, and Louise Carter. *The Gardens of Emily Dickinson*. Cambridge, Mass.: Harvard University Press, 2004.

Foley, Kevin. "Here's Some of the Best Grass in PC Gaming (And Where to Get It)." *PC Invasion*, April 20, 2021. Accessed April 8, 2023. https://www.pcinvasion.com/best-grass-pc-gaming-graphics/.

Gewin, Virginia. "The Trees' Lessons: Climate Records Are Written in Tree Rings." *Nature* 612, no. 7938 (2022): 182. https://doi.org/10.1038/d41586-022-04150-0.

Gontier, Nathalie. "Depicting the Tree of Life: The Philosophical and Historical Roots of Evolutionary Tree Diagrams." *Evolution: Education and Outreach* 4, no. 3 (2011): 515–38. https://doi.org/10.1007/s12052-011-0355-0.

Goodman, Nelson. *Ways of Worldmaking*. Indianapolis: Hackett, 1978.

Green, Amy M. "The Reconstruction of Morality and the Evolution of Naturalism in the Last of Us." *Games and Culture* 11, 7–8 (2016): 745–63. https://doi.org/10.1177/1555412015579489.

Guanio-Uluru, Lykke. "Analysing Plant Representation in Children's Literature: The Phyto-Analysis Map." *Children's Literature in Education*, 2021. https://doi.org/10.1007/s10583-021-09469-2.

Guerrilla Games. *Horizon Zero Dawn*. Sony Interactive Entertainment, 2017.

Hallé, Francis. *In Praise of Plants*. Portland, Or.: Timber Press, 1999.

Hawreliak, Jason. *Multimodal Semiotics and Rhetoric in Videogames*. Routledge studies in multimodality. New York, London: Routledge, 2019.

Heinimäki, Teemu J. "Technology Trees in Digital Gaming." In *Proceeding of the 16th International Academic MindTrek Conference*. Edited by Artur Lugmayr, 27–34. New York, NY, USA: ACM, 2012.

Kimmerer, Robin Wall. *Braiding Sweetgrass: Indigenous Wisdom, Scientific Knowledge and the Teachings of Plants*. Minneapolis, Minnesota: Milkweed Editions, 2013. https://ebookcentral.proquest.com/lib/kxp/detail.action?docID=1212658.

Kress, Gunther. *Multimodality: A Social Semiotic Approach to Contemporary Communication*. New York: Routledge, 2010.

Laist, Randy. "Introduction." In *Plants and Literature: Essays in Critical Plant Studies*. Edited by Randy Laist, 9–18. Critical plant studies: philosophy, literature, culture [1]. New York, NY: Editions Rodopi, 2013.

Lawrence, Anna M. "Listening to Plants: Conversations Between Critical Plant Studies and Vegetal Geography." *Progress in Human Geography* 46, no. 2 (2022): 629–51. https://doi.org/10.1177/03091325211062167.

Leroy, Thibault, Christophe Plomion, and Antoine Kremer. "Oak Symbolism in the Light of Genomics." *The New phytologist* 226, no. 4 (2020): 1012–17. https://doi.org/10.1111/nph.15987.

López-Cabarcos, M. Ángeles, Domingo Ribeiro-Soriano, and Juan Piñeiro-Chousa. "All That Glitters Is Not Gold. The Rise of Gaming in the COVID-19 Pandemic." *Journal of Innovation & Knowledge* 5, no. 4 (2020): 289–96. https://doi.org/10.1016/j.jik.2020.10.004.

Magnet, Shoshana. "Playing at Colonization." *Journal of Communication Inquiry* 30, no. 2 (2006): 142–62.

Meier, Sid, and Firaxis Games. *Civilization IV*. 2K, 2005.

Meier, Sid, and MicroProse. *Civilization*. MicroProse, 1991.

Mojang Studios. *Minecraft*. Mojang Studios; Sony Computer Entertainment, 2011.

Morris, Dave, and Leo Hartas. *Strategy Games*. Game guru 1. Lewes, England: Ilex, 2004.

Murray, Janet. *Hamlet on the Holodeck: The Future of Narrative in Cyberspace*. New York: Free Press, 1997.

Naughty Dog. *The Last of Us*. Sony Computer Entertainment, 2013.

Nintendo. *The Legend of Zelda*. Nintendo, 1986.

Nintendo. *The Legend of Zelda: Breath of the Wild*. Nintendo, 2017.

Nintendo. *Animal Crossing: New Horizons*. Nintendo, 2020.

Nitsche, Michael. *Video Game Spaces: Image, Play, and Structure in 3D Game Worlds*. The MIT Press. Cambridge, Mass.: MIT Press, 2008.

Nitzke, Solvejg, and Helga G. Braunbeck. "Arboreal Imaginaries. An Introduction to the Shared Cultures of Trees and Humans." *Green Letters* 25, no. 4 (2021): 341–55. https://doi.org/10.1080/14688417.2021.2072633.

Piper, Andrew. *Book Was There: Reading in Electronic Times*. Chicago, London: University of Chicago Press, 2013.

Pobłocki, Kacper. "Becoming-State. Bio-Cultural Imperialism of Sid Meier's Civilization." *Focaal—European Journal of Anthropology* 39 (2002): 163–77.

Pollan, Michael. "The Intelligent Plant." *The New Yorker*, December 16, 2013. Accessed April 8, 2023. https://www.newyorker.com/magazine/2013/12/23/the-intelligent-plant.

Roberts, Scott, and Dale Patterson. "Virtual Weather Systems." In *Proceedings of the Australasian Computer Science Week Multiconference*, 1–7. New York, NY, USA: ACM, 2017.

Schaal, Barbara. "Plants and People: Our Shared History and Future." *PLANTS, PEOPLE, PLANET* 1, no. 1 (2019): 14–19. https://doi.org/10.1002/ppp3.12.

Schulkin, Eric. "He Says It's the 'Best Grass' in Video Games, Daring Other Devs to One-up Him." *Kotaku*, March 18, 2022. Accessed April 8, 2023. https://kotaku.com/lushfoil-photography-simulator-steam-pc-tiktok-1848666248.

Segal, Howard P. "Practical Utopias: America as Techno-Fix Nation." *Utopian Studies* 28, no. 2 (2017): 231–46. https://doi.org/10.5325/utopianstudies.28.2.0231.

Strange Loop Games. *Eco*. Strange Loop Games, 2018.

Stylized Station. *Top 10 Grass in Video Games (Yes I'm Serious)*. 2022. YouTube video. Accessed August 25, 2023. https://www.youtube.com/watch?v=FDV7Jz8LexM&t=3s.

Tale of Tales. *Luxuria Superbia*. Tale of Tales, 2013.

Trewavas, Anthony. "Aspects of Plant Intelligence." *Annals of botany* 92, no. 1 (2003): 1–20. https://doi.org/10.1093/AOB/MCG101.

Wandersee, James H., and Elisabeth E. Schussler. "Preventing Plant Blindness." *The American Biology Teacher* 61, no. 2 (1999): 82–86. https://doi.org/10.2307/4450624.

Zemanek, Evi. "Pflanzliche Sinnbilder. Satirische Kritik am deutschen Wald und der deutschen Eiche als Symbole für nationale Identität und kulturelle Nachhaltigkeit." In *Literaturen Und Kulturen Des Vegetabilen. Plant Studies—Kulturwissenschaftliche Pflanzenforschung*. Edited by Urte Stobbe, Anke Kramer and Berbeli Wanning, 183–222. Studies in Literature, Culture, and the Environment v. 10. Frankfurt a.M.: Peter Lang GmbH, Internationaler Verlag der Wissenschaften, 2022.

CHAPTER 11

Theatrics of the West, and Geopolitical Economy in the Plantationocene: Opium Poppy and Its Nation-Based Phenomenological Understanding

Sindhura Dutta

Plants have existed prior to the existence of humans. However, the 18th century turn towards science and technological innovation leading to industrial revolution by the British started the massive havoc called the Plantationocene. There was nothing blasphemous about industrial revolution until it started to disseminate the idea of progress pertaining to western epistemology. Helen Tiffin and Graham Huggan in *Postcolonial Ecocriticism* has formulated their inquisition of the cardinal difficulty of what they call the environmental derangement essentially located within the "epistemological differences between European and Indian ideas of human and animal being-in-the-world" (11). This gawky difference is not limited just between the European and Indian sect but is categorical difference between the Global North and the Global South. The Global North which we know today are the countries chiefly to the West of our globe and what we know as the Global South are either warehouse countries of the west or the former colonies of England. The Global North contests its superiority on the basis of its technological privilege. The Global North approach towards the environment is exemplary of the Plantationocene epoch in which nonhuman materials of nature are rendered as passive resources for the use and abuse of human beings in the name of progress. Opium and Plantationocene have one thing in common that is Britain. The Plantationocene is an epoch that is associated with British colonialism. When the English set sail to never let the sun set on the empire, they probably unknowingly started the era of the Plantationocene. This epoch was the brainchild of the 18th-century Enlightenment, the practice of which was soon taken up by all Global North countries in the name of progress. Donna Haraway extended the concept of the Plantationocene (Haraway, 2016) that included both labour coercion culture for resource extraction and propagating the idea of plantations in Global South countries for economic benefits, for example, coercion of farmers in opium poppy plantations.

Plantations require enormous deforestation to meet the demands of agricultural commercialization and in the process, it deprives native communities

 | DOI:10.1163/9789004721777_013

and landscapes of their native plants. The Indian peasants for example had to farm opium in their lands instead of more necessary crops such as wheat and rice. They had no option of resisting the colonial insistence to plant a cash crop such as the opium poppy. British Government's goal was to progress economically; in South Asia on the backs of Indian peasants while impoverishing them and the Qing Dynasty on the backs of the Chinese addicts. The imperial government conducted monopsony trade of the plant and legalized it for economic benefit. Without this plant, the British Government would have had a difficult time maintaining its sovereignty required to rule India. When it was over with its raj, all its colonies became part of the Global South and it was these Global South countries who had to keep up with the idea of progress that the British had propagated. This progress entailed a compulsive arrangement of technological adaptation, a culture of plantations, and capitalism; and this only served the need of most of the Global North countries. The British did little to uplift the economic conditions of the peasants who farmed opium poppy for the government and such deprivation, afflicted with debt bond only made the peasants poorer. Deane Curtin coined the term "environmental racism" (Huggan and Tiffin, 2015), that talks of the hidden layers of ecological imperialism and the consequent geopolitical subjugation. If we look into the politics of environmental racism and the network between plants and racism, we will find that "the oppression of one is connected to, and supported by, the oppression of the other" (Huggan and Tiffin, 2015) by which Curtin meant that imperialism could be conducted in several ways, one of which included subjugation of communities other than the British by exercising western forest management in a way that subjugates the other to the will of the imperial master. In the case of opium poppy and trade, Britain used the plant in two ways. One was by compelling the Indians to cultivate them and the other was by trafficking the product of the same plant to China. By using the same plant, Britain colonized both the countries of the Global South.

Cultivation of resources has certain obvious environmentally detrimental effects such as deforestation or depletion of mineral resources. The cultivation of plants that the British Government promoted was guided by their own economic interests. Therefore, embedding the idea of progress by farming a cash crop such as the opium poppy was solely disguise or mere theatrics of the empire. In an astounding research, Jonathan Lehne found that the opium poppy cultivation in British India led to "lower literacy and a lower rate of public goods provision as measured by modern census data" (2) the main reason for which was "less per capita public spending on health and education by the British administration, a lower number of schools, and a greater concentration of police officers" (Lehn, 2018). The production and sale of opium poppies were done through private institutions established by the East India Company and enforced by monopsony rights by the government. The narcotic trade was so huge that the British Government in India had a separate department of Opium. At home, Indian farmers had to planted opium poppies on the orders of the British government under strict supervision and contractual obligations. The British government, Rolf Bauer writes, captured peasant interest through indebtedness and cash advances for those who didn't have easy access to credit, and it seemingly wasn't a bad offer for farmers who were looking forward to earn more by farming a cash crop since Opium had a huge demand in Asia (165). But the money made by the farmers from its sale to the company's private institutions under weren't sufficient for the production of the crop. Stuck in monopsony the farmers were prohibited from selling opium poppy to outside market and debts they acquired in advance trapped them in the colonial web of the Plantationocene. While most debates still center around the Anthropocene, we fail to acknowledge that most environmental manoeuvres for resource extraction have been a doing of Britain through its invasion of Global South countries. This typifies the overarching ambitions of man, particularly the European man under the influence of the 18th-century Enlightenment. This anthropocentric privilege of man in what we call the Anthropocene is undoubtedly a Western concept, not applicable to the East. Jean-Paul Sartre refers to Franz's *The Wretched of the Earth* (1961) in his book *Colonialism and Neocolonialism* as he continues to say "Europe believed in its mission: it had Hellenized the Asiatics, created that new species" (Sartre, 2006), and this should be sufficient in itself to understand that if not Anthropocene, the Plantationocene was a European doing in which it taught its subject the ways of colonialization. Therefore, thinking of environmental deterioration as a feature solely of the Anthropocene is a misguided practice. What the Global South is taught as progress, is a western concept as Huggan Tiffin notes in *Postcolonial Ecocriticism*. More than that, "progress" is a mythical word synonymous with the word development that the Global North has used to colonize the Global South in the 21st century. The ambivalence and paradox around the word progress have existed from the start of imperialism till date. In this case, the British government in India used a cash crop to be grown in form of plantations to propagate the idea of progress among the Indian peasants. When today, the world blames Afghanistan for its illegal opium production, the world must simply accept that it was a way its colonial master taught it. It was neocolonialism at play in which "the national bourgeoisie steps into the shoes of the formal European settlement" (Fanon 2021, 152).

If we start from the beginning of the industrial revolution in which England set sail to colonize the Global South, we will soon see the tumbling down of the innocent mask that England puts on or at least had tried to put on while conducting business with one of the largest countries in Asia. England's intervention in China during the 19th century was not just a matter of tea-opium export-import. England found the most bizarre way to enter the country, bypassing the strict regulations China had maintained for a long time. England colonized them in such a disguised manner which towards the end could only be accepted not as an intention of conducting trade but as theatrics. Opium Poppy was England's prop in its performance of the global theatre of imperialism. China had banned tobacco and the citizens looked for alternatives, which found fulfilment in Opium. What looked like a product of a plant that England wanted to supply to a country that demanded it was a mere acting, an "empty show" (The Washington Post, 2016). It was a disguise over its real intention of colonizing the otherwise closed country. The hostile geopolitical relationship between England and China goes back to the days of the two Opium Wars. It all started with Britain who "wanted tea, and other desirables such as silk and porcelain; the Qing were happy to sell. The trade was thoroughly regulated" (Lovell, 2021). But England wanted to break this thorough regulation that China under the Manchu Qing Dynasty had so severely imposed on Britain. Britain was clearly an outsider in the country. The English were addicted to the tea cultivated in China and in turn, had to buy it with their precious silver. Julia Lovell finds that between "1780 to 1790, the combined returns of the India and China trades failed to make even a £2 million dent in the £28 million debt left over from the conquest of India" (18). By 1820 the English found a way to counteract their trade deficit to China. They realized the Chinese had developed a taste for the Opium the British were producing in India. According to Lovell, the English tea addicts spent 26.25 million dollars on imports from 1752 to 1800, they found a way to balance their trade deficit by increasing exports of opium to China, which earned them around 384 million dollars by 1856 (18). The East India Company regulated groups of private agents who controlled cultivation, sales of opium poppy, auctioned in Bengal and illegally trafficking them to China since the Company eluded directly associating themselves with the drug trade. By the time the dynasty identified the problem, China was both economically and psychoactively drained by the plant. Opium and opioids prepared from the plant has for decades occupied a central position on the global stage. It has navigated its way through the politics of the nations divided between the Global South and the Global North. The Global North, from its position of privilege, has used its technocratic privilege on the world stage to dominate over the Global South in both opium production and destruction. The English used its upper hand to colonize the Chinese through the deliberate spread of drugs, whereas the US uses its technocratic privilege to apparently undermine the Taliban finances from the sales of opium and manoeuvre the use, abuse, legitimization and illegitimization of opium to control over Global South countries by propagating a holier than thou attitude. When China lost the opium war to the British Government, it was forced to open its doors to the modern capitalist world. Prior to this China remained a closed country, chiefly with a nationalist attitude. China's defeat was not only a humiliation of its ancient institution of royalty but also a breach of religious faith. This loss led to Britain's colonization of the country in terms of religious conversion which, up till their defeat in the Opium War, China had resisted so fiercely. Lovell writes that the "incontemptible vanity" that China had up till now was effectually destroyed as soon as the "missionaries became natural allies of the smugglers" as "they docked among opium traders on the island of Lintin; they interpreted for them in exchange for passages up the coast, distributing tracts while the drug was taken onshore" (20). The euphemistic confederation of missionaries and smugglers clearly showed English theatrics in making a country as close as China full of opium addicts. From producing poppies on the backs of Indian farmers, trapping them in a network of inescapable debt, to manufacturing a country (China) of addicts through drug trafficking, England capitalized its imperial experience to subjugate the Global South, as we call it today, to the penultimate extent. Once British missionaries gained access to the country, they led to the massive conversion of the Chinese to Christianity. Not only was the illegal trafficking of Opium conceived as an apparently innocent approach towards globalization and "progression" in terms of the Global North, but can but was also a methodical subjugation of Chinese nationalism, the collapse of the ancient system of emperorship that is the Manchu Qing Dynasty and the downfall of the country's economic system. The economic system collapsed in two ways. One was by draining the country of its silver, which the British used to buy tea for themselves. This meant that the British simply bought their favourite tea with the money (silver) they acquired by trafficking opium illegally into China. Secondly, by making an otherwise sober country dependent on opium, Britain destabilized the Chinese population.

The word theatrics tends to have a diverse range of meanings, mostly meaning anything related to theatre or performance. If we transcend its conventional, theatre-oriented meaning to the geopolitical stage in which I consider England and America the greatest and most timeless artists, we have to look at the dimensions of acting performed by the two biggest countries of the Global North as something cynical. Theatrical attitudes both for and against opium poppy is exemplary of both the country's hypocritic stance towards the crop depending on which country grows them and for whom. The apparent dramatic response of inhibiting opium sales in Afghanistan through airstrikes by the

United States but failing to decrease such production, or legitimizing sales of opium in China and preaching its own racial superiority (British Imperialism) is nothing but hypocrisy of the western superpowers. Opium is a plant of the colonizers that England commodified in the 19th century. It was not criminalized till the end of the empire but became a reverse trend in the 21st century, carried forward to the US who emerged as the neocolonial colonizer of the Global South. Opium Poppy production in countries like Afghanistan calls for addressing two problems; criticism of the global supplier of these drugs that finances terrorism through black market operations, and the theatrics of the West, its invasion of these countries and its failure. The United States belongs to the Global North; therefore, the American counter-terrorism measures are nothing but western theatrics of the Global North to gain dominance and respect in the East or the Global South. Their geopolitical theatrics involve implementation of emotions and policies which serve as props to these actors.

Opium cultivation in Afghanistan is a unique combination of manipulation of war conflicts, political instability and poverty. Where most villagers would, prior to the spike in demand for opium, cultivate food crops like wheat or barley, this had, however, changed since the active prominence of Al Qaeda and Taliban. Villagers are rather encouraged to farm opium, which is later sold to heroin laboratories outside the country.

The London School of Economics and Political Science published research regarding the aerial strikes the USFOR-A (United States Forces-Afghanistan) had launched in 2017 over the heroin labs in Afghanistan against a steep rise in Afghan opium production. They used the best weapons to target the buildings and laboratories with the utmost accuracy. However, crackdowns could barely administer heroin production in Afghanistan, showing a "negligible effect on the Taliban's finances" (Mansfield, 2019) through destruction of such laboratories presumed to finance the Talibans by the sales of Opium. Opium production in 2017 did decline, but that was not because of airstrikes on laboratories, as USFOR-A claimed. These laboratories were not even proper buildings but makeshift structures. The decline in opium production occurred due to the drought in the north of Afghanistan and it is amusing to note that the US spent approximately 9 billion dollars on counter-narcotics in Afghanistan, with least achieved results in the decline of the same. Nelson Pressley's article "Notes on political theater: The perils of spectacle" in The Washington Post enlightens us with the inherent hollowness of the technocratic show put up by the Global North and the West, both names used synonymously to include countries like England and the United States. Pressley associates political theatre with an empty show "likely to be signaling no genuine idea but certainly indicating a power play" (2016). It is a known fact that the west uses its technocratic apparatus to neo-colonize the east to serve its political and economic interests (Huggan and Tiffin, 29). What the west calls development is a manipulated truth that serves "different needs of those who use it, and shot through with self-congratulation and condescension" (29) and a "self-privileging discourse of neocolonialism" (32). The west takes pride in its spectacular technocratic apparatus for example when England started to colonize the world it flaunted its naval inventions whereas when America terrorized the terrorist camps in Afghanistan it did so by using best of its planes and artilleries to tear down the makeshift laboratories where the Afghans prepared Opium. The technocratic apparatus remains the west's prop in the performance of its theatrics, beside opium poppy. We may not know what the world benefited from such empty shows in the words of Pressley but we still were awestruck by its technological spectacle and the international media glamorized such precision of the American theatrics for days.

Plants have evolved for their progenitive purpose and we have coevolved with them, the common purpose being survival. The milky white latex or opium scrapped from the pods to obtain morphine, or heroin is the plant's defense mechanism to protect itself. The sap does not ooze out by itself. The farmer must pierce the sides of the pods vertically to make them bleed opium. Morphine is the exclusive feature of the opium poppy which we sow and reap. Now here I have mentioned two possibilities for experience-based approaches. We may look at the compulsive bleeding of poppies using a *nushtur* (a tapping tool) which is used to scratch the poppy pods as painful or we may see the morphine produced from the same sap as an analgesic. Dan Zahavi notes that to conduct a phenomenological reading of an object, "it is possible for one and the same object to appear in a variety of different ways" (10). Each experience of an object cannot be positioned in a vacuum that is "the perceiver to be spatially located is for the perceiver to be embodied" (Zahavi, 2019) which "calls for and involves bodily engagement and interaction" (12). They are part of a social construct, thus this unique differentiation in the experience of the same object in different ways. Another tenet in the phenomenological school of thought includes Maurice Merleau Ponty's idea of phenomenology. "To see is always to see from somewhere" (Ponty 2012) which is a definite yet distinct "gaze" which is not beyond Leibniz's "le géométral" (95). Papaver Somniferu is an object of nature. It does have le géométral but to put this opium poppy for a phenomenological analysis would bring in with it individual experience with the plant, at times dependent on the geo-spatial location of the perceiver. Her gaze will be dependent on a lot of factors and will be different from another perceiver. Merleau Ponty focuses on the le géométral of an object which has two counterparts, one how an individual will experience it and the other how God will perceive it from "nowhere" (95) and from God's eye view, the plant has flowers of varied colors, it has a pod that contains seeds, it is a plant that has been used for its sedative features. This is in Leibniz's notion, a geometrical plan of the plant seen by God from "nowhere". To see the plant from "somewhere" means the different ways people see it from there space. The Papaver Somniferu (opium poppy) here would be experienced differently by a Chinese, English or an Indian. Our relationship with the plant Papaver Somniferu or Opium Poppy as we commonly call it in the Anthropocene must be split into two parts in terms of experience of the plant. In the Anthropocene, nations have colonized the opium poppy to progress economically, whose ultimate goal is to progress its nation-based race. Because the opium war marked a significant humiliating period for the Chinese empire, it became a resentful event worthy to build contempt for the West/ the Global North. The opium poppy thus became an object of contempt. Likewise, an Indian in the Eastern side of the country must have grown reading about the miserable conditions of the farmer who were coerced or debt-trapped by the British Government in India to farm Papaver Somniferum. The English however would not experience the same object as the Indian or the Chinese would. The colonizer performs environmental racism towards the colonized nation, thereby shifting the prospective of the bodily experience to nation-based experience. The colonized likewise either writes about the painful annals of such relationship it shares with a plant that the colonizer uses as a weapon or learns its master's ways in a classic way of neo-colonialism. Literatures from the nations would likewise be split into literatures of nation-based experience such as Confessions of an English Opium Eater (Britain), Ibis Triology (India) and Cherry (United States-Afghanistan).

Cherry by Nicole Walker is a classic example of the interlocutory invasion of the US government in Iraq and Afghanistan in the name of terrorist intimidation. *Cherry* is an American novel by Nicole Walker. Walker's protagonist is an unnamed heroin addict, and his addiction starts when he returns to his country after his service in the War in Iraq. I suppose, the protagonist is unnamed for a reason. He symbolizes the many veterans of the Afghanistan war who suffer from PTSD and take to analgesic drugs. Chances are that the opioid that is consumed in America comes from Afghanistan's illegal trafficking of the plant-based drug since Afghanistan is the world's most notorious producer of the same, and America has accepted its opioid epidemic, exhibiting a simple math of demand and supply. When his addiction grows severe, he starts by selling his possessions. He moves to a shabby, cramped space to accommodate his lifestyle. The most bizarre irony of the novel is that the narrator, being a war veteran, robs a bank in order to finance his addiction. It is ironic and exemplary of America as the Global North's geopolitical theatrics. This novel brings attention to the long period of war America financed in Afghanistan which the country justifies as retaliation for a failed warning to the Taliban government that is to stop its opium production which America thought financed terrorism. America invaded a country to capture the 9/11 masterminds and, in the process, they toppled the then existing government. The US had previously complained that opium production in Afghanistan was being used to finance terrorism. However, as data shows, America did very little to curb its production during its presence in Afghanistan. Simon Spedding confronts and finds the conflict in Mallory Brown's statement "the Taliban is promoting opium production to finance terrorism." (336) He continues:

> The simple facts are that opium production was high under the US influenced government of Afghanistan of the 1970s, decreased 10-fold by 2001 under the Taliban, and then increased 30-fold and more under the US to the same level as in the 1970s. (Spedding, 336)

Cherry unravels the American pretense of saving themselves and the Afghans from the Taliban regime. Walker's protagonist in spite of being an American war veteran is barely tended for his PTSD which he acquired for fighting for his country in a futile war that was nothing but a spectacle. But nonetheless he is a white American who stands in the position of privilege. When Walker's *Cherry* is read in contrast with *The Opium Prince* by Aimaq we see the darker truth of the people of Afghanistan. *Cherry* is a narrative that provokes readers to accuse Afghanistan for promoting war and narcotics but *The Opium Prince* (2022) shows us a different consideration of the same plant. Sajadi is born and brought up in the US but returns to Kabul as an American agent, on mission to destroy poppy fields in Afghanistan, particularly belonging to Taj Maleki. Taj Maleki and others live on the money they earn from the international opium sales. They bear the burden of ground water depletion, deforestation, and climate change. Living a profession that corrupts one's environment yet fills one's pockets is a trick the colonizers have taught them. Their life depends upon poppy farms but the Americans want to destroy it. Destruction of the poppies is not as cynical as America's status in the whole mission. America is one of the largest buyers of the drug and yet it wants to destroy these plantations. We also know that although America invaded Afghanistan, it barely controlled opium production during its presence in the country. Even if the Afghans discontinued farming opium, could profits from opium poppy be replaced by any other crop?

The *Sea of Poppies* by Amitav Ghosh is the first novel in his Ibis trilogy that deals with the part and parcels of the opium trade and its reciprocation among the characters of his novel. Ghosh's *Sea of Poppies* was taken up by Rolf Bauer in his book *The Peasant Production of Opium in Nineteenth-Century India* to summarize his book with a quote from Ghosh's *Sea of Poppies*: "the English sahibs would allow little else to be planted; their agents would go from home to home, forcing cash advances on the farmers, making them sign asámi contracts" (Ghosh 2008). The British government actively financed and supported opium production for economic benefits. Neither did they care about the tiring labour that the farmers were reluctant to indulge in nor did they care about the resource depletion such production involved. The government's outright exploitation of the Indian peasants or farmers was nothing extraordinary but a culture of labour coercion that is inherently a part of the Plantationocene. It was a mere "agricultural commercialisation that many rural zones in the Global South underwent" (Bauer 2019, 5). Instead, the profits earned by the government from opium sales in China were largely used for infrastructural developments during their rule in South Asia and these infrastructures did not necessarily facilitate the Indians but the British. The Victoria Memorial is a major example of what Amitav Ghosh pointed out, sitting in the Memorial during a literary meet in Kolkata. He pointed out that "Opium was the economic foundation of Calcutta in the 18th and early 19th centuries and much of the money that went into building the Victoria Memorial came from the trade of opium" (The Telegraph, 2011). Carl A. Trocki pointed out the relevance of the plant in the business that was British Imperialism in his book *Opium, Empire and the Global Political Economy*. Trocki continues to enumerate that opium business was the sole basis of the administration of the colonial empire since "it paid the bills and provided a regular source of high-quality revenue" (10). Secondly, it created "a concentration of capitalists, in Canton and Hong Kong, in Calcutta and Bombay and it was this concentration, whose influence buttressed the imperial lobby throughout the nineteenth century" (10).

The Global South countries such as India or Afghanistan are mere factories for the Global North whose ensuing environmental crisis is not something that the superpowers are concerned about. Britain sold opium to China to fund its infrastructure and imperial rule in India through legitimized opium production in parts of Northern and Eastern India. Afghanistan too sells opium that is produced illegally due to the supervision and the policies imposed by the Global North. The American opioid abuse is nothing less than the 19th Chinese addiction to opium, and thus its scale compelled the US government to term it as an opioid epidemic, and the money Afghanistan acquired is supposedly used to fund the Taliban just like the British used the money to finance their imperial rule. Afghanistan's opium farms are illegal since it is not part of the Global North. Its opium production is not justified since it sells the opium to the Global North countries but such production and sale would be legitimate if economic development occurred in the Global North and the drug epidemic occurred in the Global South.

Something enormous may seem small depending on where you view it. The phenomenological experience of an object changes from country to country put in the historical context, the losses it resulted in, and the victories are enabled. That is why Thomas De Quincey's *Confessions of an English Opium-Eater* is not a competent exposition of the experience of neither the plant nor its Intoxicating features. De Quincey spends pages glorifying the delusional effects of the drug and his appreciation of art, but his criticism of the same is an isolated account of an English author that doesn't show the stealthy geopolitical exposition behind the opium trade of 19th century under imperialism in Asia. De Quincey's book was one of the first elaborate discussions on opium but his experience of the plant did very little to expose the network of drug trafficking or labour coercion his own country participated in during their rule in India. *Confessions* is outright racist in a tone, quite evident from De Quincey's description of the "sallow and bilious skin of the Malay, enameled or veneered with mahogany, by marine air, his small, fierce, restless eyes, thin lips, slavish gestures and adorations" (63–64) whose appearance generated bouts of anxiety in De Quincey. Contrary to *Confessions*, Ghosh's Ibis trilogy casts light on the different characters of the same East and their predicament. We already know about the different methods the British used to coerce peasants into opium production. Amitav Ghosh's protagonist Neel Ratan in *Sea of Poppies* is one of the many indentured Indians caught in the crossfire of Anglo Chinese Opium War. The transformation of Ghosh's protagonist from a Raja to Munshi alongside being an absconding fugitive on false charges framed against him and his experience of the same object as De Quincey's cannot be the same. The plant that makes De Quincey famous is the one that makes Neel a criminal. Neel too belongs to the same region of the Malay whom De Quincey condemns based on his appearance.

Bibliography

Aimaq, Jasmine. *The Opium Prince*. New York: Soho Press, 2022.

Bauer, Rolf. *The Peasant Production of Opium in Nineteenth-Century India*. Leiden, Netherlands: Brill, 2019.

Fanon, Frantz, Richard Philcox, Homi K. Bhabha, Jean-Paul Sartre, and Cornel West. *The Wretched of the Earth*. New York, New York: Grove Press, 2021.

Haraway, Donna. *Staying with the Trouble: Making Kin in the Chthulucene*. Durham: Duke University Press, 2016.

Huggan, Graham, and Helen Tiffin. *Postcolonial Ecocriticism: Literature, Animals, Environment*. London: Routledge, 2015.

Lehne, Jonathan. "An Opium Curse? The Long-Run Economic Consequences of Narcotics Cultivation in British India," June 10, 2018. http://barrett.dyson.cornell.edu/NEUDC/paper_364.pdf.

Lovell, Julia. *The Opium War: Drugs, Dreams and the Making of China*. New York, NY: Abrams Press, 2021.

Mansfield, David. "Denying Revenue or Wasting Money? Assessing the Impact of the Air Campaign Against 'Drugs Labs' in Afghanistan," 2019. https://www.lse.ac.uk/united-states/Assets/Documents/mansfield-april-update.pdf.

Merleau-Ponty, Maurice, Donald A. Landes, Taylor Carman, and Claude Lefort. *Phenomenology of Perception*. Abingdon, Oxon: Routledge, 2012.

Pressley, Nelson. "Notes on Political Theater: The Perils of Spectacle," November 18, 2016. https://www.washingtonpost.com/entertainment/theater_dance/notes-on-political-theater-the-perils-of-spectacle/2016/11/18/1c48c088-aa92-11e6-977a-1030f822fc35_story.html.

Quincey, De Thomas. *Confessions of an English Opium-Eater*. London: Vintage Books, 2013.

Sartre, Jean-Paul. *Colonialism and Neocolonialism*. London: Routledge, 2006.

Spedding, Simon J. "Is the US or the Taliban Responsible?" BMJ 336, no. 7653 (2008). https://doi.org/10.1136/bmj.39577.543333.3a.

The Telegraph. "Victoria as Opium Memorial." August 13, 2018. https://www.tel egraphindia.com/west-bengal/victoria-as-opium-memorial/cid/1277541.

Trocki, Carl A. *Opium, Empire, and the Global Political Economy: A Study of the Asian Opium Trade 1750–1950*. London: Routledge, 2005.

Zahavi, Dan. *Phenomenology: The Basics*. Abingdon, Oxon: Routledge, 2019.

Index

www.ingramcontent.com/pod-product-compliance
Lightning Source LLC
LaVergne TN
LVHW010659300926
849161LV00005B/1166